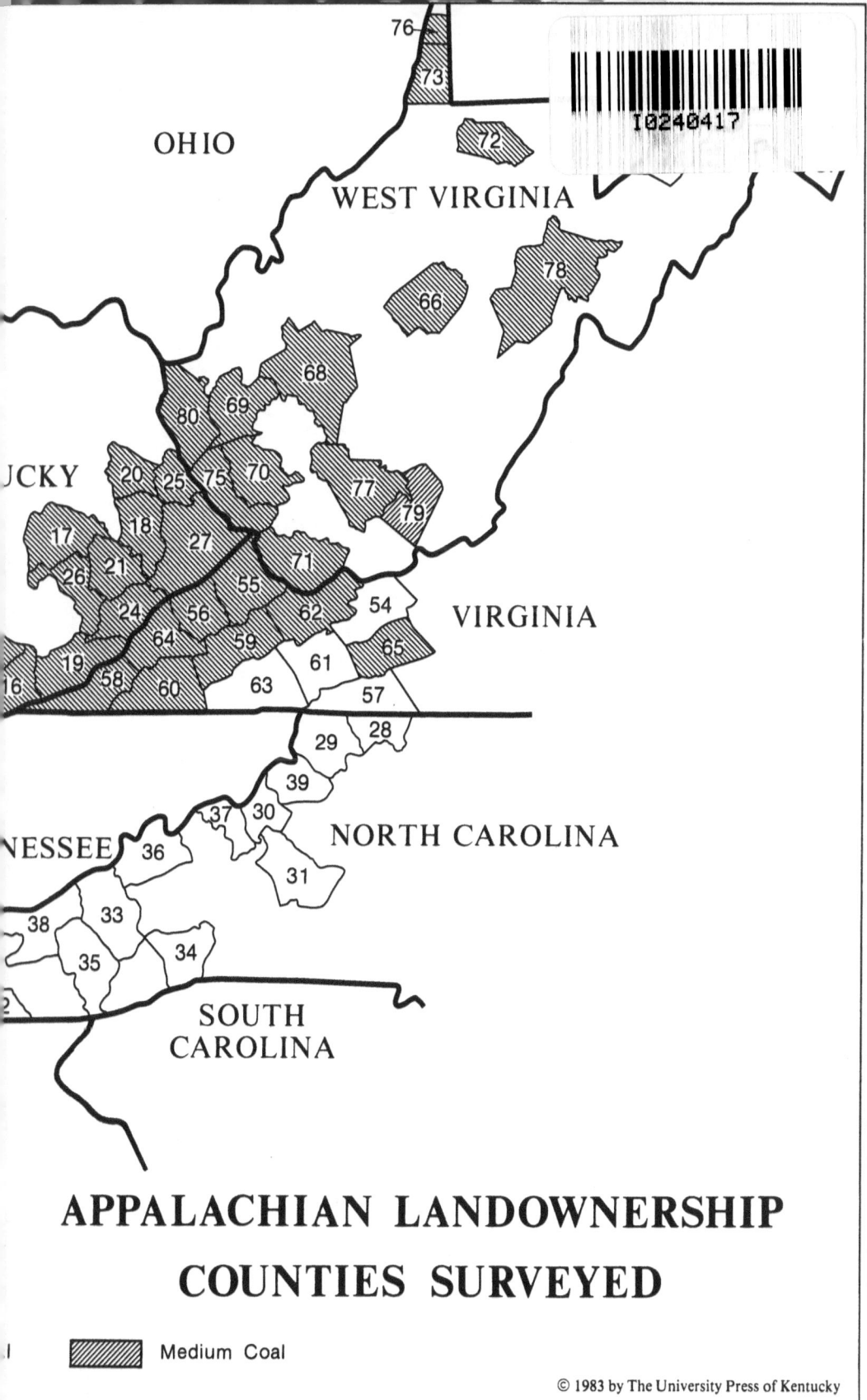

WHO OWNS APPALACHIA?

WHO OWNS APPALACHIA?

Landownership and Its Impact

The Appalachian
Land Ownership
Task Force

With an Introduction by
CHARLES C. GEISLER

THE UNIVERSITY PRESS OF KENTUCKY

Portions of this book appeared originally in the first volume of the report, *Landownership Patterns and Their Impacts on Appalachian Communities: A Survey of Eighty Counties,* in seven volumes, prepared by the Appalachian Land Ownership Task Force and submitted to the Appalachian Regional Commission in February 1981.

Copyright©1983 by The University Press of Kentucky
Introduction by Charles C. Geisler and maps
copyright©1983 by The University Press of Kentucky

Scholarly publisher for the Commonwealth,
serving Bellarmine College, Berea College, Centre
College of Kentucky, Eastern Kentucky University,
The Filson Club, Georgetown College, Kentucky
Historical Society, Kentucky State University,
Morehead State University, Murray State University,
Northern Kentucky University, Transylvania University,
University of Kentucky, University of Louisville,
and Western Kentucky University.

Editorial and Sales Offices: Lexington, Kentucky 40506-0024

Library of Congress Cataloging in Publication Data appears on page 236.

CONTENTS

List of Tables   vi

Introduction by Charles C. Geisler   ix

Preface   xxvii

Acknowledgments   xxxi

1. Landownership—A National Issue, an Appalachian Issue   1
2. Who Owns the Land and Minerals?   14
3. Who Bears the Tax Burden?   41
4. Economic Development for Whom?   64
5. Appalachia's Disappearing Farmland   80
6. Homeless in the Mountains   95
7. Ownership, Energy, and the Land   113
8. A Call to Action   136

Appendix 1. Fifty Top Owners and Other Data   149

Appendix 2. Methodology of the Land Study   157

Appendix 3. Annotated Bibliography   178

Notes   218

Index   231

LIST OF TABLES

Table 1. Surface Acres, by Type of Owner   15
Table 2. Mineral Acres, by Type of Owner   16
Table 3. Concentration of Ownership: Surface Acres   17
Table 4. Concentration of Ownership: Mineral Acres   18
Table 5. Concentration of Ownership: Most Concentrated and Most Dispersed Counties   19
Table 6. Absentee Ownership of Surface and Mineral Acres, by State   21
Table 7. Counties with Greater than Fifty Percent Absentee Ownership of Surface Acres   22
Table 8. Ownership Patterns, by Type of County and Owner   24
Table 9. Counties with Major Corporate Ownership of Surface Acres   24
Table 10. Counties with Major Corporate Ownership of Mineral Acres   26
Table 11. Surface and Mineral Holdings of Top Fifty Private Owners, by Type of Business Activity   28
Table 12. Government Ownership of Land   36
Table 13. Counties with Major Government and Private Nonprofit Ownership   37
Table 14. Counties with High Percentage of Local, Individual Holdings   39
Table 15. Legal Basis for Assessed Value of Realty, by State   42
Table 16. Property Taxes Paid Per Acre of Surface Land, by State and by Type and Residence of Owner   44
Table 17. Taxes Paid Per Surface Acre, by Size of Owner's Holdings   45
Table 18. Taxes per Surface Acre, by Land Use (*highest to lowest*)   47
Table 19. Current Value of Income Stream on One Ton of Coal Per Year, Increasing over Fifty Years   54
Table 20. Land Owned and Property Taxes Paid, by Owners   58

Table 21.	Per Capita County Expenditures, by State and by Sample Appalachian Counties	60
Table 22.	Loss of Farms and Farmland in Eighty Appalachian Counties	83
Table 23.	Correlation of Ownership with Agricultural Use of Land	85
Table 24.	Correlation of Ownership with Agricultural Sales	86
Table 25.	Tourism/Second Home Counties with High Loss of Farmland, 1969–1974	87
Table 26.	Coal Counties with High Loss of Farmland, 1969–1974	90
Table 27.	Correlation of Unavailable Land with Overcrowded Housing in Seventy-two Rural Counties	100
Table 28.	Impact of Control of Surface and Mineral Rights on Overcrowded Housing in Fourteen Tennessee Counties	103
Table A-1.	Definitions of Types of Counties	149
Table A-2.	Fifty Top Surface Owners in Eighty Appalachian Counties	150
Table A-3.	Fifty Top Mineral Owners in Eighty Appalachian Counties	153
Table A-4.	Summary of Aggregate Data Collected for Eighty Appalachian Counties	156

CHARLES C. GEISLER

Introduction
The New Lay of the Land

The great majority of Americans would, if asked, be at a loss to answer the three following questions: Who is the largest private land owner in their state of residence? How does one discover elemental information on market value, property taxes paid for a given parcel of land and real owner identity? Finally, is it true or false that patterns of landownership pervasively influence the quality of life in the community in which they reside?[1] With each passing generation, Americans know less and less about the land, its ownership and control. Even less are we aware of how this yawning ignorance affects our lives and fortunes.

Sociologists tell us that the United States is a society of strangers, a "rootless" society, a profoundly alienated society. Certainly there are many sources of this social malaise. But one profound source of social alienation which most observers underestimate is that Americans have become *physically* alienated from their land (literally, to become a foreigner on one's land). A dwindling number of Americans own real property or make their livings from it. Few today have a sustained, personal acquaintance with the land that feeds, houses, provides energy, and otherwise sustains them. The preacher's final pronouncement, "from dust to dust," is for many their only encounter with a resource their recent ancestors depended on and experienced on a daily basis.

To think that we can sever our ties with the earth is an illusion of modern invention. Since colonial times and the early Independence era, land has been at the center of the American character, the American dream, and in some instances, the American nightmare. Consider the significance of our earliest social contracts—the Declaration of Independence and the Constitution—in establishing the bond between democracy and broad-based landownership. The former enthroned the pursuit of private property as a natural right. The latter construed ownership, once attained, as inalienable. No citizen was to be physically separated from legally held land except for public purposes and unless fair compensation was made.

Little wonder, then, that Alexis de Tocqueville, traveling in the United States in the 1830s and composing his prescient work, *Democracy in America,* observed that nowhere in the world do "the majority of people display less inclination for those principles which threaten to alter in whatever manner, the laws of property." At the time of Tocqueville's visit to America, government records indicate that roughly a million acres of public lands had been surveyed and listed for sale as part of the homesteading philosophy that was to mark nineteenth-century America. Today, over a billion acres of land in the original public domain have passed into private ownership or have been granted to the states, producing a landed empire with few parallels in history.[2]

So until very recently, access to land flowed in the bloodstream of the average American and formed the basic tissue of the social body. Social status depended on the land one owned. Political institutions were shaped and reshaped by land interests—first the gentry, then the yeoman farmer, more recently the embattled taxpayer. Land has been the bosom of American culture and the birthplaace of the economy. Even today, though our agrarian vestments have nearly vanished, the export of agricultural harvests is a safety net protecting the American dollar. It is hardly surprising, then, that land and land issues have been at the heart of American reform movements across generations.[3]

The occasion for this volume is the appearance of what will be long remembered as a textbook on citizen-initiated land reform. In recent years, Appalachian residents of many descriptions have shared an intuition regarding the rural impoverishment for which their region is known. It is that land—its distorted ownership and related abuse—is something most reform legislation has ignored. This intuition spilled boldly into the public eye in 1977 when, in the wake of severe flooding on West Virginia's Tug River, thousands of people were left homeless. Regional land abuses which intensified the flooding such as forced inhabitance of the flood plain and the inability of government to find alternative homesites for the victims were manifestly traceable to the monopoly of local land by coal companies.

The distress call issuing from the Tug River floods met with coal company indifference: the directors of these companies and their stockholders lived elsewhere. Hundreds of area residents, frustrated with such complacency, felt otherwise. This group, christened the Appalachian Alliance at its 1977 meeting in Williamstown, West Virginia, vowed to focus public attention on landownership distribution in the region and on how that distribution was eroding the metabolism of Appalachian community and culture. From the group emerged a citizen's Task Force on Land. With guidance from the Highlander Center of New Market, Tennessee, and the Center for Appalachian Studies of Appalachian State University in Boone, North Carolina, the Task Force compiled the wealth of data on which this book is based.

Further help was to come from an unexpected quarter. At roughly the same time, the federally funded Appalachian Regional Commission (ARC) announced plans to conduct a rural land study of its own, the first in its fifteen-year history. Learning of the determined, grass-roots effort to accomplish a similar objective, however, ARC redirected its support to a two-year Task Force study of land and mineral ownership in the region. Signaling

Introduction

its faith in the Task Force's competence and leadership, ARC committed $130,000 to the investigation of ownership issues in six Appalachian states: West Virginia, Virginia, Kentucky, Tennessee, North Carolina, and Alabama.

The ensuing study was ambitious in scope. Under Task Force supervision, a total of 55,000 parcels in eighty counties were researched in county courthouses across these states. This represented an astonishing 20 million acres of land and mineral rights. Simultaneously, in each county over one hundred socioeconomic indicators were systematically gathered (e.g., available housing, median income, and service expenditures) in order to examine the association between ownership characteristics and community well-being. Neither before 1979 nor since has a joint venture among lay citizens, area scholars, and regional government of equal scope and diversity been undertaken. Few other analyses of the relationship between landownership and community have been so methodical, thorough, or participatory.

Here, in brief, is what the researchers set out to accomplish. In each of the six chosen states, lists were made of the top land and mineral owners. Each owner's property taxes were noted for comparison with those of local residents. These county "profiles" were then enriched with interview materials gathered from local officials, landholders and residents. In nineteen of the eighty counties, extended case studies were executed with the help of additional public records and interviews, thus permitting the unique land-community histories in roughly three counties of each state to unfold. Workshops were held, recording forms prepared, and interviewing instructions implemented to maintain uniformity in the research activities of the one hundred lay researchers gathering the data.

Before moving to a series of evaluative remarks about the Task Force results, I wish to offer unalloyed praise for what the Task Force has undertaken. Unusual fortitude and perservance were required to pursue title and tax information where records were incomplete, difficult to locate, or confounded by multiple and often secretive ownership interests. Unusual patience and vision were necessary to coordinate dozens of people with different levels of "land literacy" and to overcome the obstacles of Appalachian geography. Nor should the fear factor be discounted. As with the people who marched or rode freedom buses in the Civil Rights years or who faced derision and abuse on the antiwar picket lines of more recent years, personal risk and sacrifice were involved. Anyone who witnessed Barbara Kopple's film, *Harlan County, USA,* can fleetingly sense the courage entailed in public interest research which threatens an established power base. We should not, in a word, take the considerable effort invested in this book lightly.

The Task Force report is prescriptive as well as descriptive. Chapter 8 ventures a number of worthwhile recommendations and strategic considerations, one of which entails a national census of landownership. This proposal, the authors note, is shared by the American Institute of Planners and came close to implementation in the mid-1930's on a state-by-state basis with support from the National Planning Board. Had such documentation survived and been updated, the Task Force's current work would have been streamlined and greatly enriched with comparative information with which to document crucial ownership trends over the past half century.

Such needed data are conspicuously absent today in a systematic, comprehensive form. An important exception is agricultural landownership information in the U.S. Census of Agriculture going back to 1880 for officially recognized farms. Here, land is classified by tenure category and size of holding. As required, however, individual owner identity is protected. In the 1920 Agricultural Census, certain additional ownership information was collected, and in the late 1940s the U.S. Department of Agriculture (USDA) substantially expanded the background profile for 1945 Census landowners (Inman and Fippin, 1949). Most other landownership research from 1900 to 1950 focused obsessively on agricultural tenure categories as part of a protracted debate over rural mobility along the "agricultural ladder" and is generously summarized in the annotated bibliography by Hannay and colleagues (1949).

By default, USDA continued to guide landownership research in more recent decades, usually with a focus on farmlands. In addition to the Agricultural Census of 1954 and a cooperative effort between the Bureau of the Census and USDA's Agricultural Research Service in 1956,[4] USDA's researchers probed regional tenure patterns in the southeast (Strohbehn, 1963; USDA, 1965) and in the eastern Great Plains (Boxley, 1964). By the late 1960s, USDA was giving serious attention to a unified land data system for rural lands (Moyer, 1969). This discussion has been echoed in numerous non-USDA task forces and agencies in the 1960s and 1970s, making automated multipurpose land information systems increasingly likely.[5]

The 1970s were in many ways a turning point in landownership research for both farm and nonfarm land. USDA staff undertook landownership research at the local courthouse level, where owner identity could be traced, in West Virginia (Wunderlich, 1975) and in the northeast (Moyer and Daugherty, 1976). Further impetus to landownership studies came in the wake of the energy crisis, as foreigners, flush with fortune, broadened their investments in American real estate. That research, assisted through passage of the International Investment Survey Act of 1976,[6] coincided with evident USDA concern over the paucity of publicly available information linking landownership to wealth, land use and development (Wunderlich, 1972) and over the serious inadequacy of data describing landowner characteristics generally (Boxley, 1977).

Just as earlier Census of Agriculture research warned of notably unequal distributions of landownership in U.S. subregions (e.g., Strohbehn, 1963), so too did more recent analyses. The Agriculture Foreign Investment Disclosure Act of 1978, for example, revealed several surprises. Though foreign entities were found to own a mere 0.6 percent of total U.S. agricultural land, over 60 percent of this "foreign" land was in fact the property

Introduction

of U.S. corporations in which foreign shareholders owned a 5 percent or greater interest (Frey, 1982). In a separate analysis of foreign ownership undertaken by the U.S. General Accounting Office (GAO, 1979), an appraisal was made of investments by nonalien local and nonlocal purchasers as well. Over a third of the purchases were made by nonlocal interests of U.S. origin whereas foreign and nonlocal businesses purchased roughly one quarter (Lapping and Lecko, 1982). Findings of this kind led the U.S. General Accounting Office to conclude that the pattern of nonlocal ownership in U.S. agricultural land warranted as much attention as foreign ownership per se (GAO, 1979).

Yet another advance in publicly-sponsored landownership work emerged in 1978 with the release of figures from USDA's National Landownership Study (LOS). The results, which showed a remarkable hoarding of three-fourths of the nation's private land resources by the top 5 percent of all landowners (Lewis, 1978), came from a sample of 37,000 property owners drawn from the Soil Conservation Service's 1977 National Resource Inventory. These LOS data have been scrutinized at both the regional and state level and, together with the roughly equivalent USDA effort in the late 1940s, constitute the most ambitious national ownership inventory to date.[7] Certain other federal data sources bear on landownership, such as the 10-K Reports filed by corporations with the Security and Exchange Commission (Lewis, 1980) and the U.S. Census of Governments real estate market data base (Behrens, 1982), and have the advantage of being frequently updated.

The findings of the Task Force are generally substantiated by the data sets just discussed. So too, are they echoed in the conclusions of numerous public interest ownership studies completed in recent years. This relatively new research area relies heavily on local public records and, unlike census accounts, is frequently concerned with owner identity. This identity (i.e., public versus private, foreign versus national, corporate versus individual, etc.) is pertinent in efforts to understand environmental degradation and numerous social traumas affecting human communities. In general, the objective of such investigations is social reform, and the fine points of statistical inference to larger populations are deemed secondary.

Public interest studies of this kind are byproducts of an American land reform movement and neo-populist spirit rekindled since the early 1970s (Barnes, 1975; Geisler and Popper, 1982). Challenged by the question "Who Owns America?" researchers have pondered public records in the courthouses of Arizona, Illinois, Iowa, Kansas, Nebraska, Maine, California, and elsewhere. Summaries of these studies and their relevance to communities appear in Popper (1976), Vogeler (1981), and Walter and Stoltzfus (1982). In states such as Hawaii, New York, and Florida, publicly subsidized investigations of landownership have provided citizens groups with authoritative statements on the extent to which land holdings are monopolized or widely held.[8]

This interest has been further nourished, as it was within the Task Force, by the recent contributions of journalists (e.g., Longsworth, 1978;

Meyer, 1979); by scholars and lawyers concerned with land losses among racial minorities (e.g., FmHAa, 1981; FmHAb, 1981; FmHAc, 1981) among women (e.g., Salamon and Keim, 1979; Geisler et al., 1982), among family farmers (e.g., Martin and McLeary, 1966; Kansas Farm Project, n.d.; Guyer, 1975; Davis, 1977 and Smith et al., 1978), and among the inhabitants of mining communities and regions (Miller and Baisden, 1974; Klafehu, 1975; Illinois South, 1976; Gedicks et al., 1982). The Task Force's work will particularly enrich this final category.

Finally, of inestimable importance is the new bumper crop of citizen guides for researching disparate land records in diverse settings. The Task Force's Methodology Guide (Appendix Two in this volume), ranks among the best of these. Kindred documents include contributions by the Community Research and Publications Group (1972), the Center for Rural Affairs (1973), the Center for Investigative Reporting (Noyes, 1979), the California Institute for Rural Studies (Villarejo, 1980), the Institute for Community Economics (ICE, 1982), the Center for Urban Economic Development (1982), and the Iowa Farmers Union (n.d.). Lesser known resources for examining major urban areas are Sanborn maps, the ownership atlases of the Town and Country Publishing Company, and the plat mapping companies serving those states comprising the original Northwest Territory.[9] Consulting companies can periodically be located which maintain and market rosters of property owners in specific regions.[10]

Not long after the appearance of the Task Force's research results in 1981, the Appalachian Regional Commission convened a panel of government and academic land specialists to review the findings.[11] Reviewers were positive about most aspects of the study and tended to agree with its conclusions. Despite quarrels over methodology and tone, they praised the study's scope, the variety of information it produced, and ARC's sponsorship of such research. They noted the care used by the Task Force authors in describing their procedures and in demonstrating the feasibility of research done largely by local citizens. Appearing below is a summary of an extended memorandum submitted to ARC by the panel chairman upon completion of its review:

> The Study does make good on its most prominent claims. It shows in intimate, vivid and continuing detail what the Appalachian ownership patterns are and what they are doing to Appalachians. It provides one of the two best data sources on ownership patterns, its only possible rival being the Agriculture Department's 1978 national Landownership Survey. It also provides the best *regional* data source, supplanting the 1973 Ralph Nader report on California, *Politics of Land*. It far surpasses these other sources in its documentation of land ownership's social consequences.
>
> In addition, the Study demonstrates the feasibility of doing complex grass-roots, citizen-based ownership research in a way that will yield both regional and local information and implications. The decentralized data base and conduct of the Study must

Introduction

have presented any number of administrative difficulties, but the outcome shows that they can be overcome. As a citizen and as a researcher, I hope this successful demonstration leads to similar projects elsewhere.

I was particularly impressed by several overall features of the Study. The literature search is outstanding, and uncovered some items I had missed after five years of following ownership research. The Study, especially in its main report, is repeatedly imaginative in its use of basic data to prove complicated, often subtle points. The sheer extent of ownership concentration found is breathtaking, as are the malignity of some of its consequences and the magnitude of some of its future threats. Yet with the exception of some of the state reports, the authors avoid concluding that concentration is the root of all Appalachian evil—apparently because they are secure with their finding that it is a probable contributing cause of much of it. The conclusion that mineral ownership is more concentrated, and its taxation more inequitable, than that of surface ownership is original and striking.

But in fulfilling their charge, the reviewers also criticized the Task Force's results. These criticisms, it should be noted, tend to be directed at the various state reports and case studies, which, due to their extensiveness, are not reproduced in the present volume. Two or more reviewers, for example, made the following arguments in their summary memorandum to ARC under the general themes of methodological bias, reporting inconsistencies, and unfounded cause-and-effect. The preponderance of criticism was directed to methodological bias.

Despite its breadth and coverage, the reviewers observed, selection of central Appalachian counties was not random for either the study as a whole or the case studies. Yet inferences were frequently made to the states and the region as though the sampling was random. As the appendix on methodology indicates (page 163) each state group used its own, often differing, selection criteria. Moreover, the degree of concentration and absenteeism may in part have been artifacts of the double standard used in defining "local" interests (over 250 acres) on the one hand, and corporate, government or absentee on the other (over 20 acres). This dual standard tended to inflate the latter landowner category relative to the former.

Related to this methodological bias, the reviewers noted, was interview bias and periodic excesses of a political-moral nature in the case studies and state reports. Moreover, not all landowners were represented in the interviewing. Where corporate viewpoints were omitted, an opportunity may have been lost to show the extent of outside political and economic power being exerted at the local level. Often, "bigness" and "badness" were equated without due mention of the offsetting contributions of large interests to the region, public and private. Such problems as tone and occasional political bias are especially unfortunate, according to the ARC memorandum, insofar as they appear in the most technically proficient state studies (e.g., West Virginia, Virginia, and Kentucky).

The unevenness across state reports, just noted, is one form of inconsistency. More fundamentally, the reviewers contended that individual state reports were not consistent in themselves (across counties) or in some ways with the other state reports. This may suggest management difficulties in the overall study, caused by appreciable local autonomy at the state level. Consequently, the full state reports were not always well integrated, at times contradicted each other, and were periodically subject to factual error. Examples supporting these claims did not appear in the ARC memorandum.

The memorandum further draws attention to the study's occasional treatment of correlation (two events occurring together) as cause (one event generating another), despite rather clear statements of intent by Task Force authors (e.g., page 177). In the main, ARC reviewers felt the study rested its case on correlations at the county level where much socioeconomic data was systematically collected, but drew causal inferences about these associations. Nor were competing causal hypotheses, available to Task Force researchers, carefully disqualified. Such steps would have strengthened their arguments. This is somewhat ironic given the generous sample drawn and the computerized storage of the data—twin conditions permitting a more statistically controlled and causally reliable analysis. In this context, for example, the reviewers found the tax analysis original but somewhat oversimplified. Competing (nontax) influences on the quality of community services were not sufficiently considered.

Other reviewer criticisms could be elaborated, but are shared by most field research investigations regardless of scope. They are stubbornly present even in the absence of an agenda calling for broad citizen involvement. The reviewers were unreservedly sympathetic, despite their criticisms, to the inherent handicaps of analyzing landownership with the highly imperfect data sources available in most county courthouses throughout the United States. But perhaps more to the point, there are rebuttals to various reviewer criticisms which remove much of the tarnish they impart to the state studies and, by extension, to the summary information contained in this volume. Permit me, briefly, to bring certain additional perspectives to bear.

One notes with interest, for example, that no lay citizens were included on the review panel. Had this been otherwise, the evaluative criteria might have been broader and more sensitive to indigenous Appalachian research needs. Whenever scholars work jointly with community residents—as every scholar with first-person field research experience knows—research ideals are often modified (not necessarily fatally) in the interest of securing local trust and cooperation. A lay reviewer from Appalachia might have focused less on statistical canons and more on the limitations of performing any credible research in a region where distrust of government agencies and universities, cooptation of local residents, and intimidation by landlords of the coalfields are legendary. It is, in sum, difficult to overstate the prodigious obstacles confronting the investigation of topics as sensitive as land and mineral ownership in Appalachia.

Introduction   xvii

As for the charge of interview bias associated with spirited citizen participation, the best and ultimately only decisive test of such assertions is a restudy of the area under more controlled circumstances. Such a restudy would doubtless be welcomed by the Task Force and might absolve the original study of such allegations. Nonrandom though the survey was, it may very well have been redeemed itself by including 34 percent of all counties in the six-state region. A purely random survey strategy is perhaps of greater statistical urgency in smaller survey efforts (of which other land-ownership task forces with lower budgets should take note). But such surveys will not necessarily tap landowner realities in Appalachia with greater accuracy than the one at hand.

The dual selection standard distinguishing local from nonlocal owners is a possible source of bias. But recall that the Task Force took as its principal charge the analysis of specific kinds of ownership—absentee and corporate—rather than all ownership. Thus, its account is accurate given its intentions. Moreover, it may well be true that such 'bias' provides an accurate portrayal of the overall distribution of ownership in Appalachia. This is true to the extent that the dual standard counterbalances a serious source of bias in the other direction—the secrecy and obscurity surrounding corporate and absentee holdings endemic in the county courthouses of Appalachia. As for political overkill and moralizing in the state reports, they are, as the reviewers claim, unfortunate distractions in certain state reports but are not evident in this overall summary of the survey.

There are methodologial virtues to the Task Force's research design, unsung by the reviewers. It is commendably a multi-method approach (courthouse surveys and case studies). Even with political/moral bias, the study's results are much the stronger as the product of more than one mode of inquiry. In a limited fashion, the study aspires to use data from more than one point in time, thus reducing the static nature of its findings. Moreover, as the reviewers recognize, there is a welcome forthrightness to any study which takes pains, as this one does, to clarify its data collection methods and intentions.

Confusion over cause and correlation occurs at many levels. Although the Task Force errs in the direction of overstating causality, two things should be recalled. First, it is not too late for scholars and policymakers concerned with the Task Force's policy recommendations to perform more closely controlled analyses which might investigate the causal assertions between, say, absenteeism or ownership concentration and community deterioration, and to make these findings public. Indeed, further use of the data, collected at public expense, is to be encouraged. Second, correlation is by no means a denial of cause. In overstating their causal case, the Task Force may sacrifice credibility, but it is not necessarily creating an untruth. We should recall that the reviewers found much in the study with which to agree.

To summarize, it is certainly fair to conclude that the Task Force's research is, when all is considered, among the best in a troubled research tradition. As one of the review members observed, the study results pretend no purity and were not, in the final analysis, prepared for an academic audience. "Accepted for what it is," that reviewer concludes, "the Study is an extremely original, useful and often pathbreaking achievement."

One of the express wishes of the Task Force authors was that their labors serve as a model which other coalitions of citizens and technical researchers might emulate. This they initially accomplished by completing their work on schedule and within their budget. In numerous additional ways they have set a navigable course for others to follow with channel markers of their own strenuous experience. Yet their future work in Appalachia and that of others in their wake may be stymied by the cost of undertaking such research, even on a reduced scale. Recall that the Task Force was the beneficiary of $130,000 in federal funds. To a large degree, their example may be lost on others for want of similar good fortune. For this reason, I conclude this introductory essay with a modest proposal for overcoming this barrier. In so doing, I return, as a point of departure, to the endorsement given by the Task Force authors to the notion of a land-ownership census.

Accurate information on who owns land in America is, as noted in a recent commentary by the National Academy of Sciences (NAS,1981), a matter of considerable public interest. Systematically collected ownership data would clearly serve many groups—for example, land use and utility planners, prospective farmers and groups eager to buy land for preservation purposes, claimants and judges contending with clouded land titles, land reformers of many persuasions and surely taxing jurisdictions which, in Appalachia and elsewhere, lose considerable revenues through imprecise ownership documentation.[12] For some, land reform constitutes the conversion of private lands to public ownership in order to redistribute ownership or use rights more equitably in society. Thus specified, land reform is a radical departure from much American tradition. There is far more precedent for making public the identity and distribution of who owns the land. If, as the beginning of this essay suggests, land and landownership are deeply planted in the American character and institutional framework, then concealment of ownership and control pertaining to this essential resource is a significant hindrance to the fitness of our society and democracy.

It is perplexing to consider how much is spent in the United States on land-related information and how little—from the standpoint of reliable ownership information—the public gets in return. A recent study in Wisconsin estimated that a public investment of $78 million was made for land records maintenance and management, broadly defined (Niemann et al., 1980). The same authors reason that, if their state is roughly comparable to others, the annual expenditure could exceed $3 billion annually for the country as a whole. This is, of course, only the public expense. As every private property owner knows, transfer of land title in most of the United States is delayed by the vendor's attempts to prove ownership of the property being sold. Related title searches and insurance are costly and guarantee very little. One legal scholar, in calling for a "land reform" in U.S. land

Introduction

title registration, likens land transfer in the United States to "an aboriginal, ritualistic clambake" (McDougal, 1940). Today, this dubious service has grown more costly, while public ownership records have changed little since the last century and are ill-adapted for coping with emerging forms of ownership and lack dependability for either public interest or scholarly research.

This is not to say that a land ownership census would be problem free. In a census, owner anonymity would be protected and thereby frustrate the objectives of the Task Force and other public interest groups researching ownership matters. Perhaps more troubling, this information would be centrally stored and administered, certainly an issue for those concerned that already too much personal information is assembled in Washington. There is, however, an alternative, one more analogous to a decentralized "land library" system than to a centralized national census.[13] As with local libraries where citizens freely take out materials of interest, a local land library would permit residents to enhance their land literacy with regard to who owns and controls land assets in their community as well as other vital aspects of this resource. Currently, local jurisdictions across the country are experimenting with land libraries of various specifications and capabilities. Some make use of microcomputer and computer mapping technologies, others are assembled and maintained by hand.[14] Examples include Forsyth County, North Carolina; Lane County, Oregon; Fairfax County, Virginia; Racine County, Wisconsin; Hennepin County, Minnesota; Nassau County, New York, Wyndot County, Kansas; Montgomery County, Pennsylvania;, Nashville and Davidson Counties, Tennessee; and cities such as Chicago, Milwaukee, Virginia Beach, Houston, and Santa Rosa (NAS, 1980). North Carolina, one of the states studied by the Task Force, is today a pioneer in multiple purpose land library documentation, with nearly 40 of its 100 counties aggressively participating in that state's Land Records Management Program.

As with other reforms, this particular variant of domestic land reform at the local level is plagued with obstacles and detractors. These are political, institutional, technical and economic, and have been capably discussed by others (Costello, 1982; Bauer, 1982). Special attention might at this point be given to cost, however, since not every group wishing to perform land-ownership research will find a public agency to defray expenses. First, many of the land library experiences just cited have multiple objectives in addition to ownership information. To the extent that tax roll information is improved, as it was in Appalachia as a by-product of Task Force endeavors, less surface and mineral tax revenues are lost and more local services can be extended. Several other payment and cost recovery strategies have been proposed.[15] Currently, some of the nation's best ownership records are in the hands of the private sector, a common example being title insurance companies. Such companies could be converted to regulated utilities, that is, given monopoly prerogatives in accomplishing their objectives but subjected to greater public access and regulation.[16] Many existing utilities have excellent property maps, though they currently serve utility purposes only and tend not to be standardized across utilities. Greater standardization and public access are matters to submit to utility boards and public service commissions for approval. Costs would be added to rate structures. Finally, at least half of all U.S. communities have yet to grant access franchises to cable companies, a growth industry also dependent on reliable property owner information. Conceivably, these or other private sector interests might share the costs of land library formation, or the private sector might be contracted to produce and maintain such a system. Public assistance grants and pilot projects have periodically relieved local jurisdictions of the full cost of similar pro bono publico efforts. [17] Given the political will, there is inevitably a way to bring about such reform.

This volume epitomizes the Task Force's work. It summarizes a two-year study and serves as the capstone of an arduous research project. Its place as a model for others is well established; its findings, like the Tug River floods of 1977, are felt within and beyond the region. These pages contain a renewed sense of land as a resource for whole communities. They contain a determination wrought in Appalachian realities. And they contain the boldly portrayed accomplishment of the lay who inhabit the land.

NOTES

1. The Honorable John E. Fenton, Jr., Associate Judge of the Massachusetts Land Court, for example, recently exclaimed over "how little, both qualitatively and quantitatively, the private, and, to a lesser extent, the public sectors truly know about the land, one of the choicest natural resources." (Niemann et al., 1980:7). Fenton went on to observe that the "estrangement from full and accurate knowledge of the land diminishes our quality of life and the effectiveness of our government."

2. This figure does not include Indian reservations (see Marschner, 1959). Much land transferred to states was in turn sold to private owners in the interest of generating revenues to underwrite early infrastructure and development.

3. For a review of American land reform traditions, see Geisler and Popper (1983).

4. Findings from the 1954 Census of Agriculture appear in the 1958 USDA Yearbook of Agriculture (Wunderlich and Chryst, 1958), as do those of the cooperative study performed in 1956 (Wunderlich and Bierman, 1958).

5. Substantial research in this area have been undertaken by such diverse groups as the American Bar Association, the Council of State Governments, the American Congress of Surveying and Mapping, the U.S. Office of Management and Budget, the Departments of Housing and Urban Development and Interior and numerous other state agencies, academic and professional organizations (Niemann, 1980; NAS, 1980).

6. The resulting research is outlined bibliographically by Moyer (1981).

7. These regional studies are based on the census sub-regions of the United States as well as farm production regions (e.g., Lewis, 1978; Bills and Daugherty, 1980; Gustafson, 1980; Lewis; 1980; Moyer, 1980).

8. The New York analysis was performed by the Temporary Commission on the Adirondacks (1969) and covered twelve counties in northern New York. Florida's rural land ownership is characterized in an Agricultural Experiment Station Bulletin (Alleger, 1974) and private landholdings in Hawaii were recently summarized by the Hawaii State Land-Use Commission (1975).

Introduction

9. For example, the Rockford Map Publishers, Inc., of Rockford, Illinois, compiles ownership maps for all Wisconsin counties at the present level. For more detail on Sanborn map usage, see ICE (1982) and Wrigley (1949). These may be used in combination with Census of Housing information (see, for example, Darling, 1949) to construct portraits of rental as well as privately owned housing.

10. MacCannell and White (1982) creatively joined assessor records compiled over time by the Agri-Land Mapping Service of San Mateo, California, in determining the changing concentration of rural landownership in central California for a recent restudy of the Goldschmidt hypotheses. See also Finkler and Popper (1981).

11. This panel, called the Appalachian Land Ownership Study Implications Group, consisted of Dr. Gene Wunderlich, Dr. Robert Healy, Dr. Richard Rodefeld, Dr. Olaf Larson, and Dr. Frank Popper.

12. This list could be greatly extended; indeed, a valuable research contribution would be to list various interests in society, lay and professional, and specify their uses of ownership data. The American Public Works Association (1974) has declared, for instance, that reliable underground utility location information is a "nationwide problem." See other concerned professions in NAS (1980). High on the list of interested parties will be homeowners and prospective homeowners. The *Boston Globe* recently carried several accounts of homeowners who unknowingly bought land with major utility right-of-ways (Richard, 1982) or who, due to a surveyor's error in 1945, did not own the land their homes were built on (McCain, 1982).

13. A similar idea is advanced by Niemann et al. (1980:24), who argued that the decentralization of land records collection "will make land information more accessible and responsive. The information should be available to the citizen at the county or municipal level.... The individual could go to a place that constitutes an 'information store' and get prompt answers to questions about a particular piece of land. An information store concept also might allow user fees to support the system and to indicate the value to citizenry of the products generated by the land information system." Further support for the decentralized approach to maintaining land information comes from the National Research Council of the National Academy of Sciences (NAS, 1980:4). Mead (1981) provides a useful evaluation of some thirty statewide resource and land information systems.

14. For a discussion of how computer miniaturization and cost reduction are likely to affect land ownership research, from one's local government, office or home, see ALF (1982).

15. The ideas presented here are those of Dr. John McLaughlin of the Department of Surveying Engineering, University of New Brunswick, at the August 1982, "International Symposium on Land Information at the Local Level,' Orono, Maine. McLaughlin is a widely acknowledged authority in multipurpose cadastre development at various levels of government.

16. This notion is, upon reflection, something of a variation on the theme of converting land itself into a public utility—see Babcock and Feurer (1977).

17. For example, the U.S. Department of Housing and Urban Development gave direction in the late 1960s and 1970s to the Urban Information Systems Interagency Committee (USAC) which supported demonstration projects to develop complete or partial Integrated Municipal Information Systems in six U.S. cities (see Moyer, 1977).

REFERENCES

ALF. 1982. "The Micro-Computer and the Politics of Rural Land." The *American Land Forum Magazine* (Fall).

Alleger, Daniel E. 1974. Florida's Rural Land—How Its Ownership is Distributed.

Agricultural Experiment Station Bulletin No. 766. Gainesville: University of Florida.
American Public Works Association. 1974. Accommodation of Utility Plant within the Rights-of-Way of Urban Streets and Highways, State-of-the-Art. Special report No. 44. Chicago: APWA.
Babcock, Richard, and Duane A. Feurer. 1977. "Land as a Commodity Affected with the Public Interest." *Washington Law Review* 52 (November): 289-334.
Barnes, Peter. 1975. *The People's Land.* Emmaus, Pa.: Rodale Press.
Bauer, Kurt, W. 1982. "Problems of Political Support at the Local Level for the Development of Technically Sophisticated Land Information Systems." Pp. 11-22 in Proceedings of the International Symposium on Land Information at the Local Level (August 9-12), Orono, Maine.
Behrens, J.O. 1982. "U.S. Bureau of the Census Real Estate Market Data Base and Its Potential." Paper prepared for World Congress on Computer Assisted Valuation, Harvard Law School (August 1-6), Cambridge, Mass.
Bills, Nelson R., and Authur Daugherty. 1980. Who Owns the Land? A Preliminary Report for the Northeast States. Economics, Statistics, and Cooperatives Service Staff Paper No. 80-8, Washington, D.C.: USDA.
Boxley, R.F., Jr. 1964. Owner Characteristics and Land Ownership in the Eastern Great Plains. Economic Research Service Report No. 179. Washington, D.C.: USDA.
———. 1977. Landownership Issues in Rural America. Economic Research Service. Report No. 655. Washington, D.C.: USDA.
Center For Rural Affairs. 1973. Land Tenure Research Guide. Walthill, Nebr.: Center For Rural Affairs.
Community Research and Publications Group. 1972. People Before Property. Boston, Mass.: Community Research and Publications Group.
Costello, Michael. 1982. "The Politics and Functions of Local Government and the Land Information Process." Pp. 1-10 in Proceedings of the Inter-National Symposium on Land Information at the Local Level (August 9-12), Orono, Maine.
Darling, Philip. 1949. "A short-cut method for evaluating housing quality." *Land Economics* 25 (May):184-92.
Davis, John E. 1977. *The Corporate Connection: The Non-Farm Power in Rural Tennessee.* Nashville: Agricultural Marketing Project.
Debraal, P.T., and T.A. Majchrowicz. 1981. Foreign Ownership of U.S. Agricultural Land. Economics and Statistics Service. Agricultural Information Bulletin No. 448. Washington, D.C.: USDA
Fellmuth, R.C. 1973. *The Politics of Land.* New York: Grossman Publishers.
Finkler, Earl, and Frank Popper. 1981. "Finding Out Who Owns the Land." *Planning* 45 (August): 19-22.
FmHAa. 1981. Heir Property: Problems in the Southeast. Emergency Land Fund. Part I, The Report on Remote Claims prepared for the Farmers' Home Administration (unpublished).
FmHAb. 1981. Problems Caused by Clouded Title Resulting from Spanish Land Claims. School of Law, Natural Resources Center, University of New Mexico. Part II, Report on Remote Claims prepared for the Farmers' Home Administration (unpublished).
FmHAc. 1981. Problems Concerning Clouded Title to Land Due to Claims of Indian Tribes. Institute for the Development of Indian Law, Washington, D.C. Part III, Report on Remote Claims prepared for the Farmers' Home Administration (unpublished).
Frey, T.H. 1982. Major Uses of Land in the United States: 1978. Economic Research Service. Natural Resource Division Report. Washington, D.C.: USDA.
GAO. 1979. Foreign Ownership of U.S. Agricultural Land: A Report to Congress. Washington, D.C.: USDA.

Introduction

Gedicks, A., et al. 1982. Land Grab: The Corporate Theft of Wisconsin's Mineral Resources. Madison: Center for Alternative Mining Development Policy.

Geisler, C. C., and F. J. Popper. 1983. "Introduction." Pp. 1-26 in C. Geisler and F. Popper (eds.), *Land Reform, American Style*. Totowa, N.J.: Allanheld, Osmun, and Co.

Geisler, C. C., William Waters, and Katrina Eadie. 1982. "The Changing Structure of Female Ownership of Agricultural Land in the United States, 1946 and 1978." Paper prepared for Conference on Women and Agriculture (November), Cornell University, Ithaca, N.Y.

Giloth, Robert, Patricia Wright, and Judy Meima. 1982. Property Research for Action. Center for Urban Economic Development. Chicago: University of Illinois.

Gustafson, G.C. 1980. Who Owns the Land? A Preliminary Report for the Western States. Economics, Statistics, and Cooperatives Service Staff Report No. 80-12. Washington, D.C.: USDA.

Guyer, C. 1975. Land Ownership and Inequality in Appalachia. Unpublished thesis, Oakes College.

Hannay, A.M., D.W. Gooch, and M.S. Gould. 1953. Land Ownership: A Bibliography of Selected References. USDA Bibliographical Bulletin No. 22. Washington, D.C.: Government Printing Office.

Hawaii State Land-Use Commission. 1975. "Report to the People." Honolulu.

ICE. 1982. *The Community Land Trust Handbook*. Emmaus, Pa: Rodale Press.

Illinois South. 1976. A Handbook on Coal Leasing and Landowners Organizations. Carterville, Ill.: Illinois South Project.

Inman, B.T., and W.H. Fippin. 1949. Farm Land Ownership in the United States. Miscellaneous Publication No. 699. Washington, D.C.: USDA.

Iowa Farmers Union. n.d. Iowa Farmers Union Landownership Study. Ames: Iowa Farmers Union Office.

Johnson, B. 1974. Farmland Tenure Patterns in the United States. Agricultural Economics Report No. 249. Washington, D.C.: USDA.

Kansas Farm Project. n.d. Concentrated Land Tenure in Kansas in the 1960s. Walthill, Nebr.: Center for Rural Affairs.

Klafehn, Brad, with Mary Ann Fiske and Carolyn Ruth Johnson. 1976. The Corporate Connection. Denver: Colorado Open Space Council.

Lapping, M., and M. Lecko. 1982. "Foreign Investment in U.S. Agricultural Land—An Overview and Case Study." *American Journal of Economics and Sociology* (forthcoming).

Lewis, D.G. 1980a. "Who Owns the Land: A Preliminary Report for the Southern States." Economics, Statistics, and Cooperatives Service Staff Report 80-10. Washington, D.C.: USDA.

———. 1980b. Corporate Landholdings: An Inquiry Into a Data Source. Economics, Statistics, and Cooperatives Service Staff Report NRED 80-5. Washington, D.C.: USDA.

Lewis, J.A. 1978. Landownership in the United States. Economics, Statistics, and Cooperatives Service. Agriculture Information Bulletin No. 435. Washington, D.C.: USDA.

Longsworth, R.C. 1978. "Who Owns the Land? Well, Probably Not You and I." *Chicago Tribune* (August 6).

McCain, Nina. 1982. "Their Homes—But Whose Land?" *Boston Globe* (July 15).

MacCannell, D., and J. White. 1982. "The Social Costs of Large-Scale Agriculture: the Prospects of Land Reform in California." in C.C. Geisler and F.J. Popper (eds.) *Land Reform, American Style*. Totowa, N.J.: Allenheld, Osmun, and Co.

McDougal, Myers S. 1940. "Title Registration and Land Reform: A Reply." *University of Chicago Law Review* 8 (December): 63-77.

Marschner, F.J. 1959. Land Use and Its Patterns in the United States. Agricultural Handbook No. 153. Washington, D.C.: USDA.

Martin, J., and J. McLeary. 1966. Rural Land Ownership and Use in Tennessee. Agricultural Experiment Station Bulletin 412. Knoxville: University of Tennessee.

Mead, D. 1981. "Statewide Natural-Resource Information Systems—A Status Report." *Journal of Forestry* 79 (June): 369-71.

Meyer, Peter. 1979. "Land Rush: A Survey of America's Land." *Harper's* (January) 45-60.

Miller, T.D., and H. Baisden. 1974. "Who Owns West Virginia?" Special series to *Herald-Advertiser* and *Herald-Dispatch,* December 22-29.

Moyer, D.D. 1969. "Three Automated Land Data Systems in the United States." *Canadian Surveyor* 23 (June): 132-41.

———. 1977. "Land Parcel Systems: An Information Tool for Policy Planning and Implementation. An Information System Inputs to Policies, Plans and Programs." Proceedings of 15th Annual Conference of the Urban and Regional Information Systems Association. (Vol. III.). Chicago.

———. 1980. "Who Owns the Land? A Preliminary Report for the North Central States." Economics, Statistics, and Cooperatives Service Staff Report 80-11. Washington, D.C.: USDA.

———. 1981. Monitoring Ownership of U.S. Real Estate: A Bibliography. Economics and Statistics Service Staff Report No. AGESS810520. Washington, D.C.: USDA.

Moyer, D.D., and A.B. Daugherty. 1976. Landownership in the Northeast: A Source Book. Economic Research Service Washington, D.C.: USDA.

NAS. 1980. Need for a Multipurpose Cadastre. Washington, D.C.: National Academy Press. (Available from Cadastral Survey and Mapping Staff, Bureau of Land Management, Building #50, Denver Service Center (D-411), Denver Federal Center, Lakewood, Colorado 80225.)

———. 1981 Tracking the Nation's Real Estate: Who Owns What Land? Where? National Academy of Sciences News Report (February).

Niemann, B., Jr. (ed.) 1980. Land Record Systems Can and Should Be Modernized. Institute for Environmental Studies Report 108. Madison: University of Wisconsin.

Niemann, B., Jr., J.L. Clapp, and B.L. Larson. 1980. "The Multipurpose Cadastre: The Need, the Existing Costs, the Criteria, and the Implementation Recommendations." Pp. 5-30 in B.J. Niemann, Jr. (ed.) Land Record Systems Can and Should Be Modernized. Institute for Environmental Studies Report 108. Madison: University of Wisconsin.

Noyes, D. 1979. "Raising Hell." Center for investigative Reporting. Published by Mother Jones Magazine.

Popper, F.J. 1976. "We've Got to Dig Deeper into Who Owns the Land." *Planning* 42 (October) :17-19.

Richard, Ray. 1982. "Their Dream Home Became a Nightmare." *Boston Globe* (July 9).

Salamon, S., and A.M. Keim. 1979. "Land Ownership and Women's Power in a Midwestern farm community." *Journal of Marriage and Family* 41 (February): 109-19.

Simon, R., and R. Lesser. 1973. Land Reform and regional Ownership of Resource in Appalachia (unpublished mimeograph).

Smith, J., D. Ostendorf, and M. Schechtman. 1978. Who's Mining the Farm? Herrin, Ill.: Illinois South Project.

Strohbehn, R.W. 1963. Ownership of Rural Land in the Southeast. Economic Research Service. Washington, D.C.: USDA.

Temporary Study Commission. 1969. Final Report of the Future of the Adirondacks. Albany, N.Y.

USDA. 1965. White and Non-White Owners of Rural land in the Southeast. Economic Research Service. Washington, D.C.: USDA.

———. 1969. Land Tenure in the United States, Development and Status. Economic Research Service. Agricultural Information Bulletin, No. 338. Washington, D.C.: USDA.

Van Chantfort, E. 1982. "Land for All Reasons." *Farmline* 3 (July) : 4-8.

Introduction

Villarejo, D. 1980. Research for Action. Davis: California Institute for Rural Studies.
Vogeler, I. 1981. *The Myth of the Family Farm: Agribusiness Dominance of U.S. Agriculture*. Boulder, Col.: Westview.
Walter, R., and A. Stoltzfus. 1982. Jefferson Ignored: Land and Democracy in the United States. Washington, D.C.: Rural America.
Wrigley, Robert L., Jr. 1949. "The Sanborn Map as a Source of land Use Information for City Planning." *Land Economics* 25 (May): 216-19.
Wunderlich, G. 1972. "Who Owns America's Land: Problems in Preserving the Rural Landscape." Paper presented to the American Association for the Advancement of Science (December.)
———. 1975. Land Along the Blue Ridge. Economic Research Service. Agricultural Economics Report 299. Washington, D.C.: USDA.
Wunderlich, Gene, and Russell W. Bierman. 1958. "Farm Tenure and the Use of Land." Pp. 295-301 in *1958 Yearbook of Agriculture*. Washington, D.C.: Government Printing Office.

PREFACE

Over the past two decades community groups have been formed in Appalachia to battle such ills as the destruction of land by strip mining, the lack of land for housing, low tax base and poor services, flooding, loss of agricultural land, and broad form deeds and other unfair mineral leases. Though these are all to some extent a heritage of the landownership patterns that characterize the region, there had not until recently been a movement to deal with these patterns as underlying causes of the local problems.

In the spring of 1977 major floods hit Central Appalachia and their aftermath brought the people closer to a regional movement to change landownership patterns. The floods, whose severity was worsened by extensive strip mining in affected watersheds, left thousands homeless. Relief trailers went empty for lack of available land on which to place them, while the government refused to seize corporate land for trailer sites. Angered citizens in Mingo County, West Virginia, issued a call for groups from around Appalachia to come to Williamson to coordinate a regional response. The Appalachian Alliance, a coalition of these groups, was formed as a result. Questions of landownership were high on their agenda for study and action.

The Alliance established a Task Force on Land, later joined by interested scholars from the newly formed Appalachian Studies Conference. This coalition of community groups, scholars, and individuals became known as the Appalachian Land Ownership Task Force. It was out of this Task Force that the impetus for the study of landownership patterns arose and from which coordination of the study came.

There had never been a comprehensive study of the ownership of land and resources in the Appalachian region, nor of the related impacts of ownership patterns on economic and community development. The Appalachian Land Ownership Task Force thus proposed to the Appalachian Regional Commission (ARC) in the fall of 1978 to conduct a study with three general purposes:

1. To document ownership patterns of land in rural Appalachia, looking at such factors as extent of corporate own-

ership, extent of absentee ownership, extent of individual or family ownership, extent of local ownership, descriptions of principal owners, rate of change in ownership patterns, relationships between ownership and land use.
2. To investigate the impacts of these landownership patterns upon economic and social development in rural Appalachia, exploring the relationship of landownership patterns to land use, taxation structures, land availability for housing and industry, coal productivity, agricultural productivity, economic growth and stability, social development and stability.
3. To develop action-oriented policy recommendations for ARC; state, federal, and local officials; government agencies; and the public to assist them in dealing with problems relating to ownership patterns.

After several months of discussion between ARC and the Appalachian Land Ownership Task Force, the study finally began with a training session in May, 1979. Most of the field work was conducted in the summer and early fall of 1979, with most of the data analysis, writing, and production of the report completed in 1980. The final report was delivered to ARC in February, 1981.

The study drew upon three basic sources of information for its analysis of landownership and its impacts. First, county tax rolls were used to determine the primary land and mineral owners in rural unincorporated areas of eighty counties in six Appalachian states. Second, nineteen counties were selected for case studies designed to describe landownership and land use patterns in greater depth and to explore their impacts upon economic and community development. Third, in order to further examine the relationships suggested by the case studies, over 100 socioeconomic indicators were gathered for the eighty counties.

The data collected in this fashion produced a vast body of material. From the survey of landowners on the tax rolls, data were collected on over 55,000 parcels of land and minerals, representing some 20,000,000 acres. Hundreds of people were interviewed. Field notes and drafts of case studies amounted to some 1500 pages. The 100 socioeconomic variables for the eighty counties added to the mass. These data were processed, synthesized, and analyzed on four levels: for each of the eighty counties; for the portions of each state studied; for the regional sample; and for types of counties, i.e. coal countries, agricultural counties, and recreation and tourism counties. (A description of the methodology used in both the field research and the analysis is found in Appendix 2.)

This study is one of the few to attempt to explore landownership patterns and related impacts systematically over a given geographical region of the United States. It is the only one to have done so for such a broad area of the Appalachian region. Moreover, it was initiated by a team of citizens and scholars from the Appalachian region, who combined their experience of the region with in-depth research to produce the report. From

Preface

these unique features arise the important contributions and the limitations of the study.

The study has attempted to analyze the impact of ownership patterns on other aspects of economic and community development, an undertaking which has previously received even less attention than documentation of the ownership patterns themselves. The primary focus on landownership has at times limited the extent to which other factors in development, such as labor, capital, and topography, could also be analyzed. While the study finds that landownership patterns are necessary components of any local or regional strategies, it does not imply that changes in landownership alone would be sufficient to solve the region's economic and community development problems. Thus, any development strategies which ignore the structure of land and mineral ownership have little chance of success.

No less important than the findings of the study is the process by which it was undertaken. This project has been unique in that it was initiated and carried out by an independent task force of citizens and scholars within the region. The message is clear: with a little help from like-minded scholars, citizens can do most of the research themselves. From the beginning, the landownership study was viewed as a project that would integrate research, education, and action. People affected by the ownership patterns and active in response to them were to be an integral part of a research process culminating in the collection of data to be used in their local struggles around land issues.

The involvement of local citizens and activists carries with it a definite perspective about the importance and urgency of the problems being investigated. Indeed, it was these feelings of urgency and the realization that "disinterested" researchers were unlikely ever to tackle the controversial issue of landownership that gave rise to the study. The perspectives of other researchers would likely be different and would result in different interpretations of our data, a different research design, and so on. In these differences, no doubt, will lie the strengths of this study for some, and the weaknesses for others.

In sum, this study only begins the process of addressing the relationship between landownership patterns and the problems that persistently plague Appalachia's people. Further public debate and public action are needed to sort out the landownership issue and strategies to deal with it. The landownership issue must become an integral component of public policy decisions affecting the region as well as of the political debate about its future. Further research is also needed and it is here that Appalachian scholars need to take the lead.

Publication of the study in book form will provide a broader distribution of this important ownership data and should stimulate additional debate about the relationship between control of the land and mineral resources and the intractability of Appalachian problem areas identified in the study. The complete Appalachian Land Ownership Study consisted of seven volumes—a regional report and one volume for each of the six states studied. The state volumes contain summary state reports, county case studies, and statistical profiles of ownership and taxation in the counties. The six state volumes can be obtained from the Appalachian Center at Appalachian State University in Boone, North Carolina. This book is a revised version of the regional volume.

Findings of the landownership study have already served as a basis for numerous instances of action by community groups around the region, thus fulfilling one of the primary aims of the study. Publication of the regional volume fulfills one of our other aims: the creation of a widespread awareness and discussion of the importance of landownership patterns in Appalachia. The significance of the landownership study merits a large readership among scholars in Appalachia and elsewhere. We hope that a few policy makers and politicians will read it as well.

ACKNOWLEDGMENTS

Too often authors seem to lament the impossibility of acknowledging all the contributors to a given work and then proceed to acknowledge the most important ones. In the case of the landownership study, it is indeed impossible to acknowledge all contributors individually. However, since an aim of the study was to involve as many community people as possible throughout the Appalachian study area, the Land Ownership Task Force takes great pride in the number of people and organizations who contributed at one stage or another of the research project. To acknowledge all of those who participated in the study at every level of research and analysis would necessitate a list much longer than that of the research team listed below. Not to enumerate all of those individuals and organizations does not, however, diminish our gratitude to all those who helped make the study the collective undertaking that it was.

A few examples give some indication of the kind of participation that made the study a success. To the county tax assessors and their staffs who so often went out of their way to make ownership information available and to clarify information on the tax rolls, the task force extends a special thanks. One tax assessor, for example, spent several hours explaining the use of parcels of land identified by our researcher in that county. A special thanks also goes to those other county officials, agency representatives, and other individuals who provided landownership and use information not available from the tax rolls. Lastly, to those several hundred people who submitted to interviews for the county case studies, the Task Force owes a special debt of gratitude. Their insights were invaluable to our understanding of how landownership patterns affect the lives of the region's people.

Major funding for this study came from the Appalachian Regional Commission, to which the task force is grateful. Further funding was received from a number of private foundations to complete the project and disseminate the findings. Extensive in-kind contributions came from colleges, nonprofit community groups, and individuals in the region who generously donated time, office space, travel, computer processing and typing assistance to make the project possible. This enabled the Task Force to extend the study far beyond what would have been possible with the original grant.

The Land Ownership Task Force of the Appalachian Alliance coordinated the overall project. The project was administered by the Appalachian Studies Center of Appalachian State University and our gratitude goes to Pat Beaver and Ray Moretz for their excellent work in this regard. The actual research and production of the final report were coordinated from the Highlander Research and Education Center in New Market, Tennessee, by John Gaventa and Bill Horton, who also served as principal authors of this regional volume. Other contributors at the regional level were: Jan Collins, Steve Fisher, Jennie Freeman, Nina Gregg, Tom Holt, Jon Jonakin, Robert McClain, Juliet Merrifield, Sally Miller, Kit Olson, Richard Sanders, Linda Selfridge and Mike Wise. Additional consultation was provided by: Nancy Bain, David Brooks, Mike Clark, Deborah Tuck, and Jerry Williamson.

State teams, many members of which volunteered their services, and state coordinators did the planning, research, analysis, and writing for each state. In Alabama, Angie Wright was state coordinator and Robert Childers, James Kemp, Duna Norton, Dean Ratliff, and Amy Savory completed the research team. For Kentucky, Joey Childers was state coordinator while Don Askins, Greg Campbell, Linda Johnson, and Mark Middleton formed the remainder of the team. Cathy Efird was the state coordinator in North Carolina and was joined in the study by Pat Beaver, Babs Brown, Win Cherry, Judy Cornett, Dwayne Davis, Mary Lance, Myra McGinnis, Stephen Matchak, Sally Miller, Ray Moretz, Tom Plaut, Deloris Profit, Ellen Raim, Pam Tidwell, and Jeff West. The state coordinator in Tennessee was Charles Winfrey and other members of the team were: Rebecca Byrd, Barbara Kelly, Margaret Sharp, Lucille Shockley, Jim Stokely, and Susan Williams.

Turning to the Virginias, Tracy Weis coordinated the state effort in Virginia while Manila Bell, Steve Fisher, Susan Francis, Rose Gallagher, Richard Kirby, Phil Leonard, Clare McBrien, Michelle Martel, Catherine Molloy, Barbare Reheuser, Patrick Ronan, Carol Schommer, Judy Solberg, and Sandra Williams completed the team. For West Virginia, David Liden coordinated a research team made up of: Bill Abruzzi, Mary Adams, Margie Bean, Perry Bryant, Jeff Colledge, Mary Ann Colledge, Nora Conly, Kathy Cullinan, Rob Currie, Sarah Derosiers, Frank Einstein, Denise Giardina, Joe Hacala, John Holland, Tom Holt, Linda Martin, Mike McFarland, Ken Mills, Bob Noone, Martha Owen, Joe Peschel, Mary Margaret Pignone, Diane Reese, Judy Seaman, Paul Sheridan, Beth Spence, Bob Spence, Martha Spence, Janie Stanley, Nancy Stetten, Bob Wise, and Milton Zellermeyer.

Without the interest and commitment of these numerous participants, the project would not have been completed or even begun. Their shared vision of a better quality of life in Appalachia will be another small step closer to reality as a result of this study and the action it generates. It is ultimately their study and for the benefit of their communities.

WHO OWNS APPALACHIA?

ONE

Landownership—
A National Issue,
An Appalachian Issue

In a rural area, land joins capital, labor, and technology as a crucial ingredient for economic growth. The land and its resources are the underpinning for development. The ownership and use of the land affect the options available for future developments. The relationship of rural people to the land takes on a special meaning in their work, culture, and community life. "Throughout history," writes one land economist, "patterns of landownership have shaped patterns of human relations in nearly all societies."[1]

In the United States in recent years, the question Who owns the land? has been raised from a number of directions. Gene Wunderlich, an economist for the United States Department of Agriculture (USDA), describes the trend: "Many groups in recent years have been concerned about the concentration and distribution of wealth in America. This concern often involves the land. Corporate farming, ownership of property by aliens, accessibility of new single-family housing, the effects of real estate investment trusts, and the role of many large American corporations in natural resource and land development—all are phrases which recall the various forms this concern has taken over the last decade."[2]

The development of a concern with landownership represents, to some degree, a logical evolution in the nation's conceptions about the possession of land, and the rights and responsibilities that accompany it. Much of the early settlement and development of the nation's land carried with it a fierce ethos of the rights of the private property owner. Still today as one land-use scholar writes, "those who control much of our privately held land place extremely high value on individual freedom in doing with and to the land what the owner chooses, often without regard to the effects on the ecological system, neighbors, or the general public."[3] In the twentieth century, though, these laissez-faire attitudes regarding landownership have been challenged by new attitudes recognizing that the use of the land by one owner may affect the livelihood and well-being of others. A complex body of land-use regulations has evolved, seeking to balance the rights of ownership with responsibilities to the environment, the society, and to future generations who must use and live upon the soil.

More recently, debates over use of the land, and distribution of its benefits, have again led to questions about its ownership. One advocate of land reform in America has argued the essential connection between land use and land ownership: "It is ownership—and the economics that surround ownership—that determine whether land is farmed or paved, strip-mined or preserved, polluted or reclaimed. It is ownership that determines where people live and where they work. And, to a great degree, it is ownership that determines who is wealthy in America and who is poor, who exploits and who gets exploited by others."[4] Wunderlich, the USDA land economist, puts the implications of landownership even more broadly: "Land is a means for distributing and exercising power."[5]

In theory, the United States is well-endowed with enough land and resources to meet the needs of its people. Marion Clawson, a leading land-use scholar, points out that "in 1970, the average person in the United States had the products and the use of about eleven acres of land. . . . This land is owned by individuals, by groups and by governments, and it is used by various persons or groups, but all of us benefit, in one way or another, from its existence and from its productivity."[6] While all may benefit, studies suggest that some are more likely to benefit—or to control the benefits—than are others. Most of the population lives on the 2 percent of the United States that is classified as residential, and ownership of that land is widely distributed. But, according to best estimates, of all the private land in the United States, some 95 percent is owned by just 3 percent of the population.[7] Various governmental agencies own almost 42 percent of the land, including vast public lands in Alaska. As few as 568 corporations, according to a USDA study, own or control some 30.7 million acres of land, almost a quarter of all the American land in private hands. Worldwide, these same corporations control almost 2 billion acres—an area larger than Europe.[8]

In many countries, both agrarian and industrial, such concentrated ownership has led to land-reform policies aimed at redistributing the land, or at expanding control by the public sector over allocation of its benefits. Overseas, the United States government has openly supported such land-reform policies. Domestically, however, land reform as such has not emerged as a major policy issue. This prompts one student of rural development to argue, "Ironically the U.S. has been preaching the virtues of land reform to less developed countries since the end of World War II. The forces that resist land reform in Latin America and Asia are similar to the forces that have prevented it from becoming a subject of serious discussion in this country. But for better or for worse, land reform is as much a key to the elimination of rural poverty in America as it is anywhere else on the globe."[9]

In contrast to the lack of public debate on land-reform questions in the United States, land-use issues in the 1970s have aroused public and governmental concern. Increasingly, uses of the land for agriculture, energy, or recreation compete and conflict with one another. Increasingly, decisions about land uses involve more public scrutiny and regulation. There is growing consensus on the need to know, Who owns the land?

Perhaps the most volatile land-related issue in recent years has concerned agriculture. According to one source, "in the last twenty years, the nation has lost 60 percent of its farms. Ten farmers a day leave the land, and it is estimated that 200,000 to 400,000 farms will disappear for the next twenty years if present trends continue."[10] Behind this picture is both an internal restructuring of farming (especially a trend toward fewer and larger farms), and a loss of farmland to nonfarm uses. Both are associated with a changing pattern in ownership of American farmland.

There are a number of complex reasons for the changing ownership, including urban sprawl, the economics of farming, and land speculation by nonfarmers. The consequences of the changing ownership are far-reaching. They have to do with the most efficient size and location of farms for production of the nation's food supply; the social and political, as well as economic consequences of concentrated or monopoly control of food production; the environmental impacts of large-scale agriculture and farm and timber technologies; and the effects of ownership patterns on farm families and farm communities. Such questions cannot be fully explored without answering, Who owns America's farmland?

The 1974 Census of Agriculture found that almost 40 percent of all private farmland in the United States is owned by nonfarmers.[11] As yet there is no complete or satisfactory answer to the question of farmland ownership, but trends are visible, partially documented. One is that farms are increasing in size, "a trend pushed along as much by little farms becoming larger as by big farms becoming bigger."[12] Part of this change reflects the entrance by corporations and agribusiness into all phases of food production. In California, for instance, a 1970 study by the University of California Extension Service found that 3.7 million acres of California farmland was owned by forty-five corporate farms. Thus, one analyst concludes, "nearly half of the agricultural land in the state and probably three-quarters of the prime irrigated land, is owned by a tiny fraction of the population."[13] More recently, there have been widely publicized accounts of growing investments in farmland by pension funds, insurance companies, and other nonfarm investors.[14] A 1981 two-million-dollar study by the USDA found that "government policies which are aimed at helping farmers actually have hastened the trend towards bigger and fewer farms, and jeopardized the future of family ownership."[15]

Some of the most concentrated ownership of land in America is in woodland. Nationally, estimates suggest that over one-half of the forestland is owned by the federal government. Of the remaining, much is held by timber and paper corporations, the degree of such ownership varying from region to region. In New England, corporate ownership of timberland may be the most prevalent. Estimates in Maine suggest that a dozen pulp and paper companies own 52 percent of the state.[16] In upstate New York, the New York Temporary Study Commission on the Future of the Adirondacks found in 1970 that more than 50 percent of the private land studied was owned by 1 percent of the landowners, with three timber companies owning over 100,000 acres each.[17] Over half of the 67 million acres owned by the paper and pulp industry is in the South, though this represents only 18 percent of the region's total timberland.[18] Many observers expect the control of timberlands by corporations to grow in the South, as companies like Georgia-Pacific move their headquarters from the Northwest back to the region.[19]

The impact of farmland loss has been particularly dramatic for certain groups and regions of the country. Black landowners in the South have been particularly hard hit, especially because land serves as one of the most basic resources for the rural black community. "The more than 12 million acres of land in the South owned in full or in part by blacks in 1950 had declined to less than 6 million by 1969. For the same period, the number of black full or part time farmers declined from 193,000 to less than 67,000."[20] While the number of large farms has increased nationally in recent years, the proportion of these owned by blacks remains minuscule. For instance, in 1969, 12 percent of all Southern farms had sales of $20,000 or more, but only 2 percent of nonwhite farms fell into this category. There is little reason to believe that the trend has changed. While white landowners experienced considerable losses during this time, the losses were proportionately greater for black landowners.

In the late 1970s, another public concern, prompting quick congressional response, involved the purchase of farmland by foreign investors. The International Investment Act of 1977 authorized the president to "conduct a survey of the feasibility of establishing a system to monitor foreign direct investment in agriculture, rural and urban property." A subsequent survey by the USDA found the extent of foreign ownership to be less than one might have expected: less than one-half of 1 percent of American farmland was in foreign hands on 31 October 1979.[21] Twenty-five states developed some form of legislation limiting foreign investment in American farmland, but some observers question whether the matter of foreign ownership should be distinguished from the broader question of absentee ownership. A deputy assistant secretary of the State Department testified before a congressional subcommittee, "Foreign investment in farmland need not be regarded as a separate issue, distinct from the more general issue of absentee ownership in land and its effect on the viability of the U.S. farm."

Yet the survey of foreign ownership has not been matched by a similar investigation of absentee ownership with other holding patterns of United States farmland. However one feels about the direction of the trends outlined here, a fuller documentation of farmland ownership is needed before the public policy questions can be adequately explored.

Land use for agriculture (including cropland, grazing land, and timberland) still represents the largest use of rural land in America, yet increasingly important in this era of "energy crisis" is use of the land for extraction and production of energy, especially through mining coal and other energy sources. However, if little is known about ownership of agricultural lands, still less is known about energy lands in America, either their use or ownership. Marion Clawson, in his book *America's Land and Its Uses,* wrote, "Mining is an extremely important, though highly localized, use of the land about which we have very little information. Almost no source of data about land use provides information on mining as a land use."[22] In its multimillion-dollar study, the 1980 President's Coal Commission acknowledged the "land shortages" created in Appalachia, "in part attributable to coal companies, railroads, and other corporations owning much of the coal rich acreage." However, the commission stopped short of complete analysis, observing that "statistics for land ownership are often buried in inaccessible or untraceable county records."[23]

Slightly more is known of who owns American energy reserves under the land, though that is speculative. The last decade has witnessed growing national concern over the concentrated ownership of these energy resources, particularly by energy conglomerates. As early as 1967, a Federal Trade Commission study disclosed that five major oil companies had acquired coal rights to 2.5 million acres of public and private land. "As of 1970, 29 of the top 50 coal companies had become oil company subsidiaries, and oil companies were busily acquiring hundreds of thousands of acres of additional coal lands."[24] By 1980, oil and gas companies owned 41.1 percent of all privately owned coal reserves in the country, according to the President's Coal Commission. Six of the top ten national coal-reserve owners were primarily owned by large oil and gas companies.[25]

In addition to these oil and gas interests, the federal government is a major owner of the nation's coal resources. In the West, where roughly half of the nation's coal reserves are, the federal government is estimated to own 65 percent of the coal and to control, indirectly, another 20 percent.[26] Over the years, leasing policies allowing the development of these reserves by private interests have become matters of public controversy. The government has developed a "multiple use" philosophy, which attempts to balance environmental, energy, and socioeconomic considerations in the development of its lands. Currently, environmental interests are attempting to stall any further leasing, while development interests, spurred on by the "sagebrush rebellion," are demanding more private access to federal reserves. Regardless of the outcome of this debate, it is clear that whether and how these reserves are developed will have major effects on United States energy policy.

In shaping this policy, at least some public information exists on the location of the federally owned coal lands. However, in the East, and in parts of the West where federal ownership of energy reserves is not as extensive, little systematic data is available on the location of privately held energy resources, nor on the ownership of the lands above them. (In the Appalachian coalfields, in particular, there is extensive separation of mineral ownership from surface ownership.) A few studies of coal landownership have been done in the Appalachian area, but these are scattered and incomplete. In other parts of the country, even less information could be found.

One study, done outside of the Appalachian coalfields in southern Illinois,[27] examined 380,000 acres of corporately owned coal land in thirty-five Illinois counties. Of this land, 83 percent was owned by only six corporations. Over 99 percent of the total was owned by large absentee corporations. Small, independent company landholdings were found in only six counties and accounted for only 0.7 percent of the acreage studied. In general, the landownership reflected a national trend toward the takeover of energy reserves by integrated energy corporations.

Despite the lack of systematic information, the question of ownership of energy lands and reserves would seem to be important for shaping national energy policies. Concentrated ownership of reserves poses possibilities of monopoly control of energy supply, similar to those raised by concentrated control of energy production. Ownership and leasing patterns of private lands, as of federal lands, affect what can be mined, where, when, and by whom. Literature indicates that coal landownership is associated with other policy questions—the taxation of coal reserves; conflict between land use for energy or other needs, such as agriculture; the impact of ownership patterns on local economic development. It was perhaps with these issues in mind that Congress, in the National Energy Act of 1978, called for a study of the coal industry, including its landownership: "The study shall evaluate the economic and social impacts upon coal-producing counties and States of present and prospective land ownership patterns."[28] The study has not been done.

One of the fastest-growing demands for use of land in America is for recreation and tourism. Clawson observes that "compared with the land used by the 'big three' of grazing, forestry, and cropland, the total acreage of land in recreation use is small—about 40 million acres in the 48 contiguous states and less than 50 million in all 50 states. But the number of people rather directly concerned is large—perhaps more than half the population, the exact number is not known."[29] In response to this demand, two broad changes in ownership patterns are occurring, each with considerable controversy. On the one hand, more private land is transformed into public land to become more widely available for public use; on the other hand, more private land is bought for private recreation development.

The first transformation is seen as more and more lands are taken for national recreation areas, national parks, and national forests. The purchase of private land for public purposes, often carrying with it the threat of eminent domain by the government, has provoked considerable outcry from affected landowners. The growing restrictions on the use of public land, usually to protect its environmental and recreational qualities (e.g., the designation of wilderness areas in the Roadless Area Review and Evaluation II report), have angered private interests who seek to use the land for other purposes (e.g., mining or timbering.) These landownership and land-use changes have major consequences for the economies and cultures of the communities affected, including impacts on the use of land for agriculture or private development, development of tourism economies, and loss of land from the local tax base.

The second, often overlooked, effect of increased land use for recreational purposes is on the landownership patterns of private lands themselves. A 1976 study by the American Society of Planning Officials, *Subdividing Rural America: Impacts of Recreational Lot and Second Home Developments,* found that at least ten million recreational lots have been subdivided in the United States to be used as speculative investments, seasonal occupancy, or permanent occupancy.[30] The phenomenon of owning "recreational land" is widespread. "One U. S. family in 12 owns a piece of recreational property—either a vacant recreational lot or a second home." Such transformation of ownership, in turn, can have an impact on the future use of the land. The lots "can preclude alternative land uses and dictate patterns of growth for years to come." Moreover, such recreational land developments, while serving primarily the urban dweller, can have major consequences for the (usually rural) communities where they occur. These impacts are environmental (disruption of the land), economic (increased demands for local services, loss of land for agricultural or other purposes), and social (disruption of life-styles and communities).[31] As in the cases of agricultural or energy lands, the full extent of these impacts is difficult to assess without adequate knowledge of the landownership patterns that underlie them.

Landownership, then, is an important component of the debates on land use. Who owns the land affects how the land is used, and vice versa. Changes in ownership and use patterns can have dramatic consequences on the course of community growth. Yet, despite the importance of landownership, what is perhaps most abundantly evident is how little is known about who actually owns rural America. In his comprehensive article on American land, Peter Meyer summed up: "Almost everything about American land is known except who owns it. Somehow our vast mineral resources are assessed and quantified, mountains are measured, and ground cover and soil are analyzed.... The concept of land ownership is quite another story. It isn't part of American topography, and no atlas charts or maps the contours of proprietorship that play such an integral role in the shaping of the landscape."[32] Without such information, full assessment of the impacts and consequences of ownership is, almost by definition, impossible.

Ironically, it may have taken the public outcry over foreign ownership to provoke broader awareness of the need to know about domestic ownership as well. The attempts to find out the extent of foreign investment indicated to a number of officials how difficult such information is to obtain. A publication of the Farm Foundation and the USDA makes the point: "That inquiry [into foreign investment] highlighted what was well-known by persons familiar with U.S. real estate: The systems for recording, taxing and transferring land are not suitable for assembling information on the ownership of land. The technical, legal and economic features of the highly localized, individualized and land records systems in the U.S. resist the aggregation of land data. There was no simple, direct way of determining who owned America's land. Yet there was, and continues to be, a desire to know how wealth in land is distributed."[33]

The landownership questions of the nation are mirrored in the Appalachian Region, one of the most densely populated rural areas of the country. But as one scholar of the region wrote in 1970, "although many writers in Appalachia speak of the outside control of wealth, the degree and extent to which this is true has been only slightly and sporadically documented. There are no systematic, thorough studies of the land and mineral ownership of the region."[34]

During the 1970s, little of a general nature changed to alter the accuracy of this observation. However, several small, scattered studies emerged which did document the importance of the landownership question, and which provide models of methods for further study. (A summary of the methods used in these earlier studies may be found in the methodological appendix.) As in the discussion of land issues on the national level, the review of relevant literature in Appalachia involves looking at agricultural lands, coal and mineral lands, and recreation lands.

Appalachia is often thought of as the land of the small farmer. In fact, studies by the USDA in 1930 discovered that the southern regions of Appalachia had the heaviest concentration of small farms in the country.[35] Yet, despite national interest in the loss of farmland and the decline of the small farm, little systematic attention has been given to the contemporary plight of the farmer in Appalachia.

In many areas, though, farmlands are being lost, subject to the same pressures that affect farmlands nationally, as well as some particular pressures of the region. The development of coal lands, particularly where strip mining is involved, may limit the use of land for subsequent agricultural development. Pressures to sell land and/or mineral rights also may result in the loss of agricultural land. Building of pump-storage facilities or dams to produce electricity take prime agricultural bottomland, often in areas where such land is at a premium. Historically, Tennessee Valley Authority dams have flooded thousands of acres of farmland in east Tennessee. Recreational development and associated federal acquisitions have placed undue pressures on farmland in western North Carolina and southwestern Virginia. The conflict between agricultural and other land uses is enhanced by the fact that small-farm agriculture in Appalachia is viewed by many as economically nonviable.

Despite the general knowledge of these pressures, few specific studies have been done on the changing ownership of farmland in Appalachia, or on its related impacts on the development of the region. An exception is the study on southern Ohio by Nancy Bain and associates. They discovered a "shift away from agricultural land use... [which] declined by 56.2 percent from 1900 to 1970."[36] Accompanying the trend was the loss of resident farm owners and movement toward absentee ownership, much of it for personal or recreational purposes.

In turn, the patterns of absentee ownership have had a marked impact on the development of the area. Few of the nonresident owners have made any "improvements of the land or structures since purchasing them. The majority of parcels—60 percent—had no or an uninhabitable structure."[37] As a result of the lack of development, the absentee-owned land contributed litle to the local tax base. As one of Bain's associates summarized, "The relative disuse of absentee land may ... impede the region's agricultural development as well as property taxes."[38]

The quality of development in a rural agricultural community may be affected by the size of ownership, as well as by absentee ownership (as was found in the California study by Walter Goldschmidt).[39] In Alabama, students at the University of Alabama compared the ten counties in the state having the smallest-average-size farms with the ten counties having the largest-average-size farms, in terms of agricultural productivity, land-use tenure patterns, and indicators of community development. Almost every indicator of economic and social well-being was more favorable in the small-farm counties. For example, the small-farm counties had twice as much revenue from ad valorem taxes and over two and a half times as much total tax revenues. Additionally, they had twice as many miles of county roads, and spent one-third more on education. The median income was almost twice as high, the poverty rate and proportion of substandard housing was half that of the large-farm counties. The small-farm counties were located predominantly in the Appalachian section of northern Alabama.[40]

Perhaps in no section of Appalachia have landownership and its related impacts been a greater issue than in what is known as Central Appalachia (eastern Kentucky, southern West Virginia, southwestern Virginia, portions of eastern Tennessee). It is in these areas where coal production is predominant. And it is also in these areas where a pattern of absentee corporate landownership has been verified in numerous studies, historically and today.

In much of this region, purchase of land and mineral rights by absentee, corporate interests began in earnest in the last half of the last century. Harry Caudill, one of the best-known writers of the region, describes the process in this way: "After the Civil War industrialists were able to glimpse the outlines of the nation's coming growth and they foresaw the indispensability of Appalachian coal. Agents of coal and iron companies and ambitious speculators moved in to corner title to the mineral deposits the geologists had located."[41] Throughout much of the region, a rapid change in landownership patterns occurred, often transforming small agricultural and homestead holdings to large absentee and corporate hands. The change was greatest in the Central Appalachian coalfields, though it extended to Southern Appalachian timber stands and to other resources as well. Historian Ron Eller described:

> By 1910 outlanders controlled not only the best stands of hardwood timber and the thickest seams of coal but a large percentage of the surface land in the region as well. For example, in that portion of western North Carolina which later became the Great Smoky Mountains National Park, over 75 percent of the land came under the control of thirteen corporations, and one timber company alone owned a third of the total acreage. The situation was even worse in the coalfields. According to the West Virginia State Board of Agriculture in 1900, outside capitalists owned 90 percent

of the coal in Mingo County, 90 percent of the coal in Wayne County, and 60 percent of that in Boone and McDowell counties.[42]

Since the turn of the century, the land question has arisen again and again in studies of the region. For instance, the report of the 1926 President's Coal Commission referred to the concentration of corporate ownership, observing that the U.S. Steel Corporation and its subsidiaries owned 750,000 acres of coal lands in Appalachia; Consolidation Coal owned 340,000 acres; and Pittsburgh Coal and Coke, 164,000 acres (though, the commission concluded, there were "relatively few instances where companies owned far in excess of what is needed to protect their investments.")[43] In the 1930s, Watkins, a British analyst, took a stronger position: for the development of independent communities in Appalachia, he said, "a necessary step ... would seem to be much larger and stricter control over the ownership of land, for in many cases the operating companies own all of the land within convenient reach of the mines."[44]

With the advent of the War on Poverty in the region in the 1960s, the issue of ownership of the region's land and mineral wealth again began to be raised. In every state in Central Appalachia, studies of landownership, varying in quality and scope, questioned why such poverty existed amid such land and resource richness.[45]

One of the earliest such studies was done in 1969 by Richard Kirby for the Appalachian Volunteers. Kirby began his study with the observation, "Poverty in the United States has always seemed especially cruel and ironic so close to so much bounty. In eastern Kentucky, the paradox has yet another layer of irony: some of America's poorest people live literally on top of some of America's richest land."[46] Kirby then asked, Who owns east Kentucky? and searched for an answer in county tax records of eleven east Kentucky courthouses. In answer, he found that some thirty-one people and corporations owned about four-fifths of east Kentucky's coal. About 86 percent of the coal land was owned by absentee interests. While concentrated, absentee interests controlled the wealth, they returned little in the way of property taxes to needy county coffers. About the same time, a journalist for the *St. Louis Post Dispatch* found the same pattern of undertaxation. In explanation, a Kentucky tax commissioner was quoted as saying, "the coal companies pretty much set their own assessments. ... We have no system for finding out what they own."[47]

During the same period, the theme of poverty-amid-wealth was again echoed in West Virginia. Writing in the *New Republic*, Paul Kaufman observed that "West Virginia is notorious not for the money it gets but for the money that corporations take out of it." Looking at the nine southernmost counties, Kaufman found that "nine corporations own more than one-third of the land in these counties, and the top 25 landowners control more than half. Of the nine dominant corporations, only one is a West Virginia company doing business principally within the state."[48] About the same time, a public interest research team headed by Davitt McAteer at the West Virginia University Law School surveyed the top fourteen coal-producing counties in the state and found a similar pattern: twenty-five landowners owned approximately 44 percent of the counties studied—yet payed only about one-tenth of the real-estate taxes.[49]

About five years after the McAteer study, Tom Miller, an investigative journalist for the *Huntington Herald Dispatch,* conducted a further statewide search in an attempt to answer the question, Who owns West Virginia? "Certainly not West Virginians," he found. "More than two-thirds of the non-public land in the state is controlled by outside interests. These are giant fuel, transportation and lumber companies."[50] Combining mineral and surface rights, he found the problem to be pervasive. "In almost 50 percent of West Virginia counties, at least half of the land is owned by the out-of-state corporate interests." Direct ownership of land, he found, was extended through control of land and minerals by leasing: citing a 1971 report by the West Virginia Public Service Commission, Miller said that thirteen companies leased 3.8 million acres in West Virginia, and that the amount was climbing by one-half million acres a year. The combination of ownership and leasing meant that absentee landlords "own or control two-thirds of the land in this mineral-rich state." At the same time, "they reap the benefit of low tax assessments, often paying as little as two cents per acre in annual property taxes for valuable coal, timber or oil and gas holdings."[51]

The patterns of concentrated corporate and absentee ownership of coal lands, accompanied by low tax assessments, have also been found in the Tennessee coalfields. In 1971, a study by three Vanderbilt University students of the five major coal-producing counties in northeastern Tennessee found that nine large corporations controlled 34 percent of the land surface and approximately 80 percent of the coal wealth. Yet, in 1970, they accounted for less than 4 percent of the property-tax revenue of these counties. Most of the concentrated ownership was found in the portion of these counties with the major coal reserves, while the remaining parts of the counties retained more dispersed, individual ownership.[52]

The picture in the southwestern Virginia coalfields does not change. A 1973 study there found that fifteen corporations owned 602,283 coalfield acres, accounting for from 10 percent to 69 percent of the surface of the counties studied.[53] One company alone, Pittston Coal, owned 41 percent of this acreage. A further study by Carol Schommer in 1978 documented the inadequate assessment of coal lands in southwestern Virginia. Noting the increase in the fair market value of coal over the previous ten years, she found that the assessed value of coal had not risen. As in the case of the other coalfield states, concentrated absentee ownership carried with it underassessment of mineral reserves.[54]

It must be recognized that the evidence these studies present is still incomplete. They were done by different methods at different times, and for selected counties. They do not extend to many coalfield sections of the region, such as Alabama. Though the evidence is still fragmentary, the picture is a consistent one of concentrated corporate ownership, with a great extent of absentee ownership. In his study *Poverty Amidst Riches: Why People are Poor in Appalachia,* John Wells summarizes what the studies say:

"Corporate entities own at least 4,340,142 coal-rich acres of central Appalachia. Of this total, the top five corporations have 1,594,446 acres or 37 percent; the top ten control 2,442,635 acres or 56 percent; the fifteen largest own 2,977,798 or 68 percent; the twenty majors control 3,274,770 acres, in excess of 75 percent." As for the rate of absentee ownership, more than 77 percent, some 3,357,491 acres, is held by firms located out of state. This ranges from a low of 37 percent in Tennessee to a high of 85 percent in West Virginia. In short, Wells concluded, "We have found that a small minority of mighty corporations control the wealth, and that most of these are absentee."[55]

If the coalfields of central Appalachia are associated with absentee corporate ownership, other parts of the region are affected by absenteeism of a different sort: that connected with second homes and development of the recreation and tourism trade. Two decades ago as part of a "definitive" study of Southern Appalachia, John Morris argued, "There is little reason to doubt that the potential of the tourist industry is much greater than has been realized to date, and that properly developed it will be a tremendous asset to the Appalachian economy."[56] Indeed, there has been much tourist development in areas like western North Carolina, east Tennessee, some sections of southwest Virginia, the Allegheny highlands of West Virginia, and some counties in northern Alabama. However, tourism and recreational growth has been a mixed blessing. Other studies indicate that the loss of agricultural land, inflated land prices, increased pressure on local services, dislocation, and destruction of local cultures have all been negative side effects of recreational development.[57] What little evidence there is also indicates that tourism carries with it a dramatic change in landownership patterns.

The principal study available of landownership patterns in areas heavily affected by tourism and recreation development was conducted in western North Carolina by the North Carolina Public Interest Research Group. The study found "evidence of a boom in recreational development, including second home building, that was, on the average, three times higher than the state average." With the recreation boom came a change in landownership patterns. In ten selected mountain counties, the study discovered a "remarkable" 26 percent increase in the level of nonlocal (out-of-county) ownership between 1968 and 1973. During the same period, the holdings of North Carolinans declined, while those of out-of-state holders increased by almost 50 percent. Despite the property boom, there was "very little in the way of significant economic gains for the counties" in which recreational development was taking place. Though there had been negative environmental effects, the study concluded that "effective land use planning activities by county governments in the region have been meager and superficial."[58]

Many of the areas experiencing resort and second-home growth are located near public lands, owned principally by the federal government through agencies such as the Forest Service, TVA and the U.S. Army Corps of Engineers. Much of the national debate on use and development of public lands focuses on the West, where the federal holdings are far more extensive than in the East. However, the federal government is the largest single landholder in Appalachia, and its presence is significant when viewed from the local vantage point. Si Kahn, in one of the few systematic investigations of federal lands in the region, wrote (of the Forest Service): "On the local level, the amount of National Forest Land in many counties in the Southern Mountains is staggering. Within the Appalachian areas of West Virginia, Virginia, Tennessee, Kentucky, North Carolina, and Georgia, there are 37 counties in which the Forest Service owns over 20 percent of the land. In 14 of these counties, more than 40 percent of the land is in national forests."[59] These extensive holdings have given rise over the years to widespread, and strongly put, public feelings about the federal land presence and policies. A Forest Service officer in western North Carolina describes his perception of these feelings: "There is strong resentment toward interlopers, claim jumpers, tourists, and out-of-state owners. Many are unwilling to accept outside ownership and development of what they consider their lands. Sophisticated concepts like public ownership on land they have just recently subdued, upon which they are dependent, and that have been used freely as hunting ground by their families since settlement, are not accepted."[60]

Condemnation policies, in particular, have created a residue of ill-will and distrust among some toward the Forest Service and other federal agencies. Also, since the federal holdings are tax exempt, except for what the agencies pay "in lieu" of taxes, they can greatly affect the tax base of a county and its subsequent ability to provide services. For instance, Kahn estimated in 1974 that "the Appalachian National Forests cost local governments nearly $10 million a year in lost revenues—revenues that could go to support schools, roads, health programs, welfare, and other public services."[61] Finally, public lands are seen to encourage certain developments such as recreation and second homes, or to discourage other developments, such as industry or mining, the consequences of which are valued by some and opposed by others. Whichever position one takes, an underlying point remains: federal holdings do have major impacts on communities where they are located, and yet little systematic study exists either of the extent of the federal presence or of the effects of the federal policies.

Again and again, the question of Who owns the land? emerges, be it in reference to use of the land for agriculture, for energy, or for recreation purposes. The debates on the national level over land ownership and land use are mirrored in rural Appalachia, where a number of studies have examined ownership of the region's farms, energy resources, and recreation areas. In general, we find that these studies have been localized, uneven in quality and varying in approach. Remarkably little systematic, comprehensive attention has been paid to ownership questions. There has been even less systematic investigation into the consequences of the ownership patterns. This study, then, will turn to the two-fold task of documenting ownership patterns in rural Appalachia and examining the related impacts of land ownership, particularly in the areas of property taxation and delivery of services, economic development, agriculture, housing, and energy and the environment.

TWO

Who Owns the Land and Minerals?

> My wife was named Anna Morla. She was the third daughter of a poor farmer and I was the third son of a wealthy one, and our families lived near each other in a mountain valley with a little river running through it, one deep enough for swimming, an idyllic place, and that river was our courting road, our site of poetry and dreaming. . . . And when finally we ran off and got married, my father on our return, after much lecturing in his anger, did let me have sixty rocky acres of land for my own, and did come together with others of that mountain community to build us a small house, and did lend me a plow and a hoe and an ax and a cow and an ox, so in April we took our broken things to our own land and built our first fire in our own place together.
>
> John Ehle, *Time of Drums*

The image of Appalachia as the land of rugged individuals, owning and working relatively small family holdings, is strong in the literature about the region. But unlike the young couple in Ehle's novel, today the image for so many remains a dream. The reality is a region where the ownership of land is concentrated in relatively few hands, dominated by absentee and corporate holders, with little available for local families to work, farm, or otherwise to enjoy.

For this study, data was collected on the ownership of over 20 million acres—13 million acres of surface rights and 7 million acres of mineral rights—in eighty Appalachian counties spanning six states. Using county courthouse records, the information was gathered on over 55,000 parcels of property, owned by some 33,000 owners. To the knowledge of the Land Ownership Task Force, this data bank is the largest ever collected on the ownership of Appalachia, and possibly of rural America. (Tables 1 and 2 show the number of surface and mineral acres examined in each state.) A brief sketch can give the basic picture:

Only 1 percent of the local population, along with absentee holders, corporations, and government agencies, control at least 53 percent of the total land surface in the eighty counties. This means that 99 percent of the population owns, at most, 47 percent of the land. Of the 20 million acres of land and minerals owned by over 30,000 owners in the survey, 41 percent —over 8 million acres—are held by only fifty private owners and ten government agencies.

Of the 13 million acres of surface sampled, 72 percent—almost three-quarters—were owned by absentee owners; 47 percent by out-of-state own-

Land and Minerals

Table 1. Surface Acres, by Type of Owner

	Acres Individual	Acres Corporate	Acres Government/ Private Nonprofit	Total
Alabama	2,003,106	1,260,162	313,487	3,576,755
	(56%)[a]	(35%)	(9%)	(100%)
	(28%)[b]	(18%)	(4%)	(50%)
Kentucky	708,262	665,517	208,483	1,582,262
	(45%)	(42%)	(13%)	(100%)
	(23%)	(21%)	(7%)	(51%)
North Carolina	601,579	267,761	592,087	1,461,427
	(41%)	(18%)	(41%)	(100%)
	(21%)	(9%)	(20%)	(50%)
Tennessee	1,118,457	1,041,212	281,165	2,440,834
	(46%)	(43%)	(11%)	(100%)
	(29%)	(27%)	(7%)	(63%)
Virginia	900,581	539,140	389,987	1,829,708
	(49%)	(30%)	(21%)	(100%)
	(26%)	(15%)	(11%)	(52%)
West Virginia	593,485	1,369,203	352,659	2,315,347
	(26%)	(59%)	(15%)	(100%)
	(13%)	(30%)	(8%)	(51%)
Total	5,925,470	5,142,995	2,137,868	13,206,333
	(45%)	(39%)	(16%)	(100%)
	(24%)	(21%)	(8%)	(53%)

Source: Appalachian Land Ownership Study, 1980. Using 1978–79 property tax records, this survey recorded all corporate, public, and absentee owners above 20 acres and all local individual owners above 250 acres in the unincorporated portions of the county. (The survey covered 53 Percent of the total surface of the eighty counties.)
[a] The percentage of the land sampled for each state.
[b] The percentage of the total surface in the sample counties in each state.

ers and 25 percent by owners residing out of the county of their holdings, but in the state. Four-fifths of the mineral rights in the survey are absentee owned.

Almost 40 percent of the land in the sample, and 70 percent of the mineral rights, are corporately held. Forty-six of the top fifty private owners are corporations, among them some of the largest corporations in the country. (See Tables A.2 and A.3 and "A Profile of Top Corporate Owners"). While some 45 percent of the land in the sample is owned by individuals, well over one-half of this is owned by absentee individuals. The remaining portion of the land in the sample (16 percent) is owned by

Table 2. Mineral Acres, by Type of Owner

	Acres Individual	Acres Corporate	Acres Government/ Private Nonprofit	Total
Alabama	710,839	870,073	716	1,582,528
	(45%)[a]	(55%)	(.05%)	(100%)
	(10%)[b]	(12%)	(0%)	(22%)
Kentucky	246,772	357,576	11,182	615,530[c]
	(40%)	(58%)	(2%)	(100%)
	(8%)	(11%)	(.4%)	(19%)
North Carolina	128,671	78,659	0[d]	207,330
	(62%)	(38%)	0	(100%)
	(4%)	(3%)	0	(7%)
Tennessee	202,753	435,046	0	637,799
	(32%)	(68%)	0	(100%)
	(5%)	(11%)	0	(16%)
Virginia	96,180	557,588	0	653,768
	(15%)	(85%)	0	(100%)
	(3%)	(16%)	0	(19%)
West Virginia	774,032	2,458,299	27,345	3,259,676
	(24%)	(75%)	(1%)	(100%)
	(17%)	(55%)	(1%)	(73%)
Total	2,159,247	4,758,141	39,243	6,956,631
	(31%)	(68%)	(1%)	(100%)
	(9%)	(19%)	(.2%)	(28%)

Source: Appalachian Land Ownership Study, 1980.
[a] Percentage of mineral acres sampled in state
[b] Percentage of total surface acres in sample counties in each state
[c] Does not include mineral acres in several counties which were unavailable at the time of study
[d] Adequate data on federal mineral ownership was unavailable.

government and nonprofit bodies—ten government agencies account for 97 percent of this public ownership.

For many areas of Appalachia, who owns the mineral rights is just as important as who owns the surface. Despite the fact that millions of acres of mineral rights in Appalachia are simply not recorded for tax purposes, the study discovered almost 7 million mineral acres, equal to 28 percent of the total surface area of the eighty counties. A large portion of these mineral rights is held separately from the surface land, and bought or sold as a separate commodity, consequently having major impacts on the use of the surface land.

CONCENTRATION OF OWNERSHIP

Of all of the indicators of landownership, perhaps the most significant is concentration—the degree to which land is held by relatively few owners, or the degree to which it is dispersed among the many. From other studies, one can suspect that the greater the concentration of landownership in an area, the greater the ability of a few owners to dominate the area's development; the more dispersed the ownership, the more likely that economic power will be dispersed. The extensive study of landownership in California, *The Politics of Land*, argued, for instance, that "almost by definition, highly concentrated ownership and control of land mean more political and economic power and greater ability to oppose contrary interests than do widely diffused ownership or control. Large landholders direct a greater portion of their earnings toward political ends than do smaller holders. And the large owner's land use decisions have greater public impact, thus giving him greater bargaining power with officials."[1]

In this study, measures of concentration will necessarily understate the extent of concentrated ownership actually present. First, the concentration of ownership can be given only among the owners sampled, not for all owners in a county (as this information was not collected). Second, on the aggregate level, it was not always possible to combine all parcels owned by the same owner, across all counties, because ownership might be recorded under different names (though this was attempted where possible).

Despite the methodological problems, the point stands clear: the ownership of land in Appalachia is highly concentrated in relatively few hands. The top 1 percent of the owners in the sample own 44 percent of the land in the sample—over 1,400 times what is owned by the bottom 1 percent of the owners in the sample. The top 5% own 62% of the land, contrasted to the bottom 5 percent who own .25 percent, or about 250 times less than what the top 5 percent own. The top half of the owners in the sample control

Table 3. Concentration of Ownership: Surface Acres

Owners in Sample	Percentage Surface Acres in Sample	Percentage Total Acres in Survey Counties	Concentration Index*
Top 1%	43.5	21.9	1,450
Top 5%	62.2	31.3	249
Top 25%	84.9	42.7	45
Top 50%	94.4	47.4	17
Bottom 1%	.03	.02	—
Bottom 5%	.24	.13	—
Bottom 25%	1.90	.95	—
Bottom 50%	5.60	2.82	—

*Percentage of sample owned by top x% of owners, divided by percentage of sample owned by bottom x% of owners.

94 percent of the land, the bottom half control under 6 percent. (See Table 3.)

The ownership data for minerals is less complete than the data for land. Nevertheless, the pattern of concentration remains. The top 1 percent of the recorded mineral owners control 30 percent of the mineral rights in the sample—some 15,000 times greater than what is owned by the bottom 1 percent of the mineral owners. The top 5 percent of the recorded mineral owners own 62 percent of the recorded minerals; the top 50 percent own 97 percent. (See Table 4.)

In order to make comparisons among counties and types of counties, it is possible to develop an index that measures the degree of concentration or dispersal of land and minerals among owners. For the study, several such indexes were calculated.[2] The simplest, however, is obtained by dividing the percentage of land owned by the top x percent of owners by the percentage of land owned by the bottom x percent of owners. The higher the index, the greater the concentration; the lower the index, the lower concentration. For instance, in the overall sample, the top 25 percent of the owners own 85 percent of the land; the bottom 25 percent own 1.9 percent; the index of concentration (at the 25 percent level) is 45. For the recorded mineral acres the index is 136.0.

Using this index (at the 25 percent level), one finds that land ownership is most concentrated in the counties with the highest coal reserves: In those counties the top 25 percent of the landholders own 56 times the land owned by the bottom 25 percent of the owners in the sample. This may be contrasted with the counties with no known coal reserves, where the index is 31. For counties with a high degree of tourism as the economic base, the index is 40. For the high agriculture counties, the concentration of ownership is lowest. There, the top 25 percent of the owners own 35 times that owned by the lowest 25 percent of the owners.

Table 4. Concentration of Ownership: Mineral Acres

Owners in Sample	Percentage of Mineral Acres in Sample	Mineral Acres as Percentage of Total Surface Land	Concentration Index*
Top 1%	30	9	15,000
Top 5%	62	17	1,240
Top 25%	90	25	136
Top 50%	97	27	32
Bottom 1%	.002	.0006	—
Bottom 5%	.05	.01	—
Bottom 25%	.66	.08	—
Bottom 50%	3.20	.89	—

*Percentage of sample owned by top x% of owners divided by percentage of sample owned by bottom x% of owners.

Land and Minerals

Using the index, it is also possible to identify counties where concentration is likely to be high, and thus where a few landholders are likely to be able to dominate the county's development. (See Table 5.) In six counties —Swain, N.C.; Raleigh, W.Va.; Harlan, Ky.; Kanawha, W.Va; Wise, Va.; Sequatchie, Tenn.—the index is over 100, i.e., the top 25 percent of the owners own over 100 times what the bottom 25 percent own. In twenty-eight of the eighty counties, or 35 percent, the top 25 percent of the owners own 50 times that of the bottom 25 percent of the owners. Five of the top six counties are in the coalfields, primarily with corporations as large owners. Swain County, where concentration is highest, is affected by the vast federal holdings in that county.

By no means is the concentration index as high for all of the counties surveyed. In sixteen, or 20 percent, of the counties surveyed it is under 20. In two counties—Mineral, West Virginia, and Ashe, North Carolina—it is under 10. In other words in these counties we can find a relatively equal distribution of land. Both of these counties lie outside the coalfields, have

Table 5. Concentration of Ownership: Most Concentrated and Most Dispersed Counties

Most Concentrated (top 20 counties in sample)	Concentration Index*	Most Dispersed (top 20 counties in sample)	Concentration Index*
Swain, N.C.	150	Mineral, W.Va.	9.0
Raleigh, W.Va.	135	Ashe, N.C.	9.6
Harlan, Ky.	116	Jefferson, W.Va.	11.0
Kanawha, W.Va.	115	Watauga, N.C.	11.1
Wise, Va.	108	Ohio, W.Va.	11.5
Sequatchie, Tenn.	103	Russell, Va.	11.7
McDowell, W.Va.	96	Alleghany, N.C.	11.9
Logan, W.Va.	89	Marion, W.Va.	13.7
Bell, Ky.	87	De Kalb, Ala.	14.7
Van Buren, Tenn.	86	Lincoln, W.Va.	15.8
Campbell, Tenn.	83	Scott, Va.	17.3
Scott, Tenn.	78	Blount, Ala.	19.2
Mingo, W.Va.	66	Henderson, N.C.	19.2
Mitchell, N.C.	65	Lamar, Ala.	19.5
Marion, Tenn.	62	Roane, Tenn.	19.8
Dickenson, Va.	61	Madison, N.C.	19.8
Avery, N.C.	61	Breathitt, Ky.	20.9
Braxton, W.Va.	60	Wayne, W.VA.	21.0
Anderson, Tenn.	59	Knox, Ky.	22.2
Walker, Ala.	57	Lee, Va.	22.8

*Percentage of the sample owned by top 25% of owners divided by percentage of sample owned by the bottom 25% of owners. The correlation between this measure and the more complicated Gini coefficient, which was also computed, is high: .735 at the .001 level of probability.

Source: Appalachian Land Ownership Study, 1980.

little government ownership, and are principally agricultural in base. Both, however, are seeing increasing second home and corporate buying.

ABSENTEE OWNERSHIP

Like concentration, the residence of an owner can be highly significant in determining the impact of ownership patterns in a local community. In this study, residence refers to whether an owner lives in the county, out of the county but in the state, or out of the state altogether. All owners living out of the county in which their property was located were defined as being absentee. Not only are Appalachia's land and mineral resources tightly held, they are also held primarily by absentee owners.

The extent of this absentee ownership in the region is enormous, beyond even what the previous studies of landownership in Appalachia might have suggested. Of the 33,465 owners in the survey, 81 percent, controlling 72 percent of the acreage sampled were nonlocal. Some 47 percent of the land sampled was owned by out-of-state owners: 25 percent was owned by owners living in the state but out of the county. Altogether, this absentee-owned land in the survey is equivalent to 38 percent of the total surface of the land in the eighty county area. (See Table 6.)

The pattern of absentee ownership persists, and grows stronger, when mineral rights are considered. Of the almost 7 million acres of mineral rights in the sample, 79 percent are absentee-owned—52 percent by out-of-state owners and 27 percent by in-state/out-of-county owners. Expressed in terms of the land surface in the survey area, 22 percent of the total area of the eighty counties is underlain with absentee-owned minerals (and this, it should be remembered, includes only the mineral rights that are recorded). When mineral and surface acres are combined, 15.1 million acres, or some 75 percent of the acreage surveyed, is absentee-owned

The vast majority of these absentee owners—87 percent—are relatively small owners, owning between 20 to 250 acres. However, the total acreage these small owners control is relatively low—representing only 18 percent of the absentee-owned acres in the sample. In fact, when acres controlled are examined rather than number of owners, one finds that as the holdings in Appalachia become larger and more concentrated, so also are they more likely to be absentee. Of holdings between 20 and 500 acres, 64 percent are locally held. But, of holdings above 1,000 acres, the reverse is true—75 percent of them are held by out-of-state or out-of-county owners.[3]

From previous studies of landownership in Appalachia, one might have expected absentee ownership to predominate primarily in the major coal counties. The expectation does not hold. Absentee ownership is pervasive throughout the region, regardless of the rural economic base. In fact, of the counties with no coal reserves or only minimal coal reserves, 73 percent of the land is absentee held, compared to 72 percent for the major coal counties. Outside the coalfields, absentee coal owners are replaced by very large timber companies, federal holdings, second-home owners, or recreation developers.

Table 6. Absentee Ownership of Surface and Mineral Acres, by State

	Surface Acres			Mineral Acres		
	Out-of-State Ownership	Out-of-County, In-State Ownership	Total Absentee Ownership	Out-of-State Ownership	Out-of-County, In-State Ownership	Total Absentee Ownership
Alabama	1,281,170 (36%)[a] (18%)[b]	1,147,225 (32%) (16%)	2,428,395 (68%) (34%)	605,257 (38%) (9%)	724,507 (46%) (10%)	1,329,764 (84%) (19%)
Kentucky	878,894 (56%) (28%)	363,624 (23%) (12%)	1,242,518 (79%) (40%)	342,417 (56%) (11%)	151,244 (25%) (5%)	493,661 (81%) (16%)
North Carolina	970,162 (66%) (33%)	319,338 (22%) (11%)	1,289,500 (88%) (44%)	127,705 (62%) (4%)	66,348 (32%) (2%)	194,053 (94%) (6%)
Tennessee	905,749 (37%) (23%)	788,384 (32%) (20%)	1,694,133 (69%) (43%)	329,599 (52%) (8%)	203,084 (32%) (5%)	532,683 (84%) (6%)
Virginia	991,509 (54%) (28%)	314,638 (17%) (9%)	1,306,147 (71%) (37%)	429,132 (66%) (12%)	127,483 (17%) (4%)	556,615 (83%) (16%)
West Virginia	1,206,539 (52%) (27%)	384,070 (17%) (8%)	1,590,609 (69%) (35%)	1,781,870 (55%) (40%)	632,522 (19%) (14%)	2,414,392 (74%) (54%)
Total	6,234,023 (47%) (25%)	3,317,279 (25%) (13%)	9,551,302 (72%) (38%)	3,615,980 (52%) (14%)	1,905,188 (27%) (8%)	5,521,168 (79%) (22%)

[a]Percentage of surface acres in the sample for that state.
[b]Percentage of total surface acres in the survey counties in that state.
Source: Appalachian Land Ownership Study, 1980.

In one-fourth of the counties in the study, the absentee-owned land in the sample represented over one-half of the total land surface in the county. The counties are indicative of the kinds of absenteeism found throughout the region. (See list of these counties in Table 7.) In Swain County, vast federal holdings are joined by corporate developers and second-home owners to leave little land held by local individuals: in that county, 80 percent of the land is held by the federal government. Of the remaining land, 23 percent is owned by twenty-one companies, fifteen of which are Florida-based land-development companies; and 40 percent is owned by out-of-county individuals. In the plateau counties of Sequatchie and Van Buren in Tennessee, the holdings of one timber company, J. M. Huber Corporation, account for much of the absentee-owned land. In the mountainous coal regions of McDowell and Mingo or Logan counties in West Virginia; Knott, Harlan, and Martin, Kentucky; Wise, Virginia, or Campbell, Tennessee, absentee-based coal and energy companies dominate the scene.

CORPORATE OWNERSHIP

"Somewhere we lost ourselves. I think it was when the companies bought up the land," a West Virginia farmer said. The largest, and most

Table 7. Counties with Greater than 50 Percent Absentee Ownership of Surface Acres

	Percentage of County Surface Absentee-Owned	Percentage of Sample Absentee-Owned	Acres Absentee-Owned
Swain, N.C.	94.0	99	315,139
Sequatchie, Tenn.	81.1	98	141,692
McDowell, W.Va.	79.3	94	270,647
Mingo, W.Va.	67.9	90	183,717
Van Buren, Tenn.	66.8	71	108,578
Clay, N.C.	63.6	97	85,048
Logan, W.Va.	63.0	71	149,891
Marion, Tenn.	62.9	85	203,864
Dickenson, Va.	60.6	92	128,845
Campbell, Tenn.	58.3	76	168,299
Shelby, Ala.	58.0	87	297,026
Knott, Ky.	57.6	82	131,195
Harlan, Ky.	57.6	78	172,757
Martin, Ky.	57.2	91	84,590
Bledsoe, Tenn.	56.8	75	146,946
Winston, Ala.	56.1	86	206,202
Morgan, Tenn.	55.9	81	192,926
Jackson, N.C.	55.3	89	173,700
Wise, Va.	54.6	85	143,723
Scott, Tenn.	52.6	70	181,217
Bland, Va.	51.4	73	123,080

Source: Appalachian Land Ownership Study, 1980.

Land and Minerals

likely to be absentee, of Appalachia's nongovernment owners are corporations. Altogether, corporations own 5,142,995 acres of the land surveyed, amounting to 20 percent of the land mass in the eighty counties. The corporate land is held by some 3,100 owners, with a relatively large average holding of 1,660 acres each. Of these 3,100 companies, the top 46 own 56 percent of all of the corporate land in the sample. In twenty-four of the eighty counties, corporate-owned land accounted for more than half of the surface acres surveyed.

In addition, the corporations own 4,758,141 acres of mineral rights, representing 68 percent of the mineral rights surveyed. Expressed as a percentage of the surface land in the counties, these corporate-held mineral rights underlie 19 percent of the surface. The mineral rights are held by fewer owners and in larger parcels than the surface. Only 1,100 owners own this almost 5 million acres of minerals, an average plot of 4,087 acres. Overall, in forty-six of the sixty-four counties where data on mineral wealth was recorded, corporations own over one-half of the mineral acres.

While much of Appalachia's land and mineral wealth is thus corporately owned, little of it is held by local businesses. Of the just over 5 million corporate acres in the survey, 84 percent are absentee-owned; 60 percent by out-of-state owners. For the mineral wealth of Appalachia, the relationship between corporatism and absenteeism increases. Of the 4.8 million acres of corporate-owned mineral acres in the survey, 89 percent are absentee-owned; 62 percent by out-of-state corporations.[4] These absentee corporate owners are also likely to be the larger of Appalachia's owners. Overall, forty-six of the top fifty owners in the survey are corporations—only two of them have their headquarters in the county where their major holdings are found. While the average plot of land held by locally owned corporations is only 75 acres, it is 1,400 acres for the out-of-county corporation and 2,670 acres for the out-of-state corporations.

While absentee ownership is found to be pervasive throughout the region, corporate ownership is more predominant in certain portions of the region than in others. In the "high coal" counties in the sample, 50 percent of the land in the sample is corporately held, compared to 31 percent in the high agricultural counties, and more than double the rate of corporate ownership in counties with tourism as its base. (See Table 8.) Not only do the coal counties have greater corporate ownership than the other county types, but the level of corporate ownership also increases with the level of coal reserves. In the medium coal counties, with 10 million to 100 million tons of known reserves, 31 percent of the land in the sample is owned by corporations; and in the counties without coal resources, 20 percent—only two-fifths the rate of corporate ownership in the high coal counties. The same pattern is true for mineral rights. Four-fifths of the mineral rights in the survey are found in the thirty-three counties with a high level of known coal reserves. Of these, 72 percent are corporately held.[5]

If corporate ownership of land, with its related characteristics of being absentee held and in large plots, is most likely to be extensive in counties

Table 8. Ownership Patterns, by Type of County and Owner

Type of County	Acres Individual	Acres Corporate	Acres Government/ Private Nonprofit	Total
High coal[a]	2,920,090	3,652,272	752,919	7,325,281
(N = 33)	(40%)[d]	(35%)	(10%)	(100%)
	(21%)[e]	(27%)	(6%)	(54%)
High agriculture[b]	3,109,262	1,775,043	928,402	5,812,707
(N = 30)	(53%)	(31%)	(16%)	(100%)
	(25%)	(15%)	(8%)	(48%)
High tourism[c]	1,871,352	882,717	1,098,548	3,852,617
(N = 19)	(48%)	(23%)	(29%)	(100%)
	(29%)	(14%)	(17%)	(60%)

[a] Known reserves greater than 100 million tons per county.
[b] Annual sales of over $5 million per county (based on 1974 Census of Agriculture).
[c] More than 25 percent of their service industry in tourism and recreation-oriented services (based on 1974 Census of Service Industries).
[d] Percentage of land in sample for that type of county.
[e] Percentage of total surface in counties of that type.

Table 9. Counties with Major Corporate Ownership of Surface Acres

	Percentage of County	Percentage of Sample	Corporate-Owned Acres Surface
McDowell, W.Va.	75.9	89.9	258,984
Logan, W.Va.	67.2	92.6	196,239
Raleigh, W.Va.	64.4	91.8	249,334
Mingo, W.Va.	62.5	82.6	169,228
Sequatchie, Tenn.	60.6	68.2	105,923
Campbell, Tenn.	57.5	75.3	166,000
Harlan, Tenn.	55.2	74.7	165,733
Van Buren, Tenn.	50.9	63.3	82,719
Shelby, Ala.	45.7	68.7	233,527
Wise, Va.	45.2	70.2	118,944

Source: Appalachian Land Ownership Study, 1980.

Land and Minerals

with the most coal reserves, a list of the ten most corporately held counties in the sample should come as no surprise. (See Table 9.) Four of the top five most corporately held counties are in southern West Virginia, the so-called "heart of the billion-dollar coalfields." In these four counties, almost 90 percent of the land in the sample is corporately held, accounting for over two-thirds of *all* of the land in those counties. Campbell County, Tennessee, is dominated principally by one corporate owner, Koppers Company of Pittsburgh, which owns 96,000 acres in the county, which it would like to develop for synthetic fuel production. Wise County, Virginia, and Harlan County, Kentucky, are owned by an assortment of coal landholding companies and Shelby County, Alabama, by the vast holdings of four paper companies, U.S. Steel, and Southern Railroad, which has recently merged with the Norfolk and Western. Of these ten most corporately held counties, only Van Buren and Sequatchie, Tennessee, do not appear in the list of counties with high coal reserves, though they are affected by the ownership of the J. M. Huber Corporation, a timber concern and the largest corporate holder found in the survey.

In the case of mineral rights, corporations may own several seams of minerals at varying depths. When the acreage of these seams is combined, the result is greater than 100 percent of the total surface acres of a county. Thus, in looking at the 10 counties with the highest degree of corporately held mineral rights (Table 10), one can see that in Lincoln and McDowell counties, West Virginia, corporate-owned mineral rights are equivalent to 120 percent and 105 percent, respectively, of the total land surface in each county! One can also see that eight of the ten counties with the greatest degree of corporation ownership of minerals are in West Virginia. This may be due primarily to the fact that the mapping of mineral rights for tax purposes is more extensive there than in other states. As discussed earlier, in many counties, mineral rights simply may not be reported to the assessor, or if they are, they are vastly understated. In Perry County, Kentucky, for instance, the Kentucky River Coal Company reports owning 26,272 acres of coal for tax purposes, while in actuality it owns over 75,000 acres of minerals in the county.

Regardless of the case of underreporting by corporations of their minerals, the case studies make clear that the ownership of minerals underground may strengthen and expand the corporate control gained through surface ownership. In the case of Lincoln County, West Virginia, for example, corporations own only 10 percent of the surface in the county, while they control mineral acres equivalent to 120 percent of the county's total land mass—and the county has suffered, as a consequence, the same negative impacts experienced by counties with extensive corporate domination of surface lands. Of the sixty-four counties in which mineral rights are recorded, however incompletely, corporately controlled mineral rights represent a greater degree of the county's surface than does corporately held land in twenty-six of them.

Because the ownership of minerals may extend the control of an area gained through surface ownership, the two may be combined to give a more

Table 10. Counties with Major Corporate Ownership of Mineral Acres

	Corporate Mineral Acres as % of County Surface	Corporate Mineral Acres as % of Mineral Acres Sampled	Corporate-Owned Mineral Acres
Lincoln, W.Va.	120.4	91.8	337,385
McDowell, W.Va.	104.9	86.2	357,935
Mingo, W.Va.	97.5	85.9	264,046
Marion, W.Va.	89.7	83.1	178,519
Raleigh, W.Va.	87.1	88.7	337,272
Logan, W.Va.	84.8	74.3	247,595
Marshall, W.Va.	77.7	99.1	151,219
Ohio, W.Va.	77.1	89.2	52,284
Dickenson, Va.	71.7	96.2	152,422
Buchanan, Va.	65.5	74.6	213,165
Martin, Ky.	59.6	60.2	88,070

Source: Appalachian Land Ownership Study, 1980.

complex Index of Resource Control (the percentage of surface owned plus percentage of minerals, expressed as a percentage of surface).[6] The Index for corporate ownership is 39, meaning that the combined mineral and surface ownership of corporations in the sample is equal to 39 percent of the total surface of the eighty counties. For the counties with the greatest known coal reserves, the Index rises dramatically to 56—i.e., corporate-owned surface and mineral acres are equal to well over one-half of the total land mass in these counties. In eight of the counties, the combined surface and recorded mineral acres owned by corporations is equivalent to 100 percent or more of the county's surface acres. These are McDowell, W.Va. (181); Mingo, W.Va. (161); Logan, W.Va. (152); Raleigh, W.Va. (151); Lincoln, W.Va. (130); Dickenson, Va. (115); Sequatchie, Tenn. (104); Martin, Ky. (100).

Who are these top corporate owners of Appalachia? Tables A.2 and A.3 in Appendix 1 list the fifty top nongovernmental surface and mineral owners in the survey.[7] Twenty-four of the top mineral owners are not among the large surface owners. Together, these seventy-four top private owners (the fifty surface and mineral owners and the twenty-four additional holders of minerals only) control almost one-third of the 20 million acres surveyed. Of the top fifty surface holders, forty-six are corporations, owning 2,884,569 acres—over half what is owned by the 3,100 corporations identified in the survey. Of the top fifty mineral owners, forty-two are corporations, owning 2,815,790 mineral acres or 60 percent of all the corporately held minerals in the sample.

Some of these large owners in Appalachia represent the largest and most well-known corporations in America. Others are relatively small and anonymous nationally, yet like the larger corporations they possess through their vast holdings tremendous ability to influence both the exploitation of nationally needed resources and the course of community development where their holdings are located. For this reason, public policies in Appalachia must take into consideration the plans and powers of the corporate owners of the region's land and mineral wealth. In order to do so, knowledge of who these major corporate owners are and why they are holding the resources is essential.

As can be seen in Table 11, of the top fifty surface owners, nine are wood and timber companies, owning an average of almost 100,000 acres each. The next largest owners of surface lands are companies whose principal business is coal mining or holding coal lands. Some seventeen of these coal companies own 764,333 acres, followed by steel and other metal companies (444,910 acres), oil, gas, and energy companies (294,323 acres), railroads (255,286 acres), miscellaneous corporate holders (227,559 acres), and individuals (121,753 acres).

For the mineral owners, the picture changes—oil and gas companies account for 945,375 acres of mineral rights, most of which are not oil and gas, but coal. Coal and coal land companies come next with 755,928 acres; railroads have 326,232 acres, and steel companies 317,531 acres. Timber companies, who are principal surface owners, have far fewer acres of mineral rights recorded on the books (though they may, in fact, own them).

A better understanding of these corporate holdings can be gained by looking more in depth at each corporate type. When surface and mineral acres are combined, seventeen coal-mining and coal landowners own 1,520,261 acres. The surprising characteristic of these owners is that only Pittston, Alabama By-Products, and Blue Diamond Coal Company are engaged primarily in the business of mining coal. The others simply lease their land and minerals to coal operators who do the mining.

In 1965, *Dun's Review of Modern Business* wrote of these coal land corporations, "for all their small numbers . . . these coal royalists hold what may be one of the most lucrative investments in all of America."[8] The "coal royalists," as they are called, simply oversee their land (usually through a local manager) negotiate leases, and collect the royalties, currently as high as $2.00 to $3.00 per ton. The companies who lease the land for the mining incur most of the risks.

On the national economic scene, these coal landholding companies are small and often relatively unknown. Even their trade group, the National Coal Lessers' Association is not highly visible. Yet, locally, these companies are often viewed as having enormous power. Through single decisions of their offices, the land use of huge portions of certain counties can be affected. Coal operators are dependent upon good relations with them to negotiate the leaseholds necessary to mine the coal, which often provides the jobs in an area. Tenants living in old coal camps on their property may also be dependent upon these companies' good will for housing. Whole communities are potentially affected by the taxes and economic base which their resources provide.

Table 11. Surface and Mineral Holdings of Top Fifty Private Owners, by Type of Business Activity

Business Activity	Surface Acres	Mineral Acres	Total Acres (Surface + Mineral)
Coal and coal lands	764,333 (25.4%)[a] (17)[b]	755,928 (24.4%) (14)	1,520,261 (24.9%) (19)
Oil, gas, other energy	294,323 (9.8%) (6)	945,375 (30.5%) (8)	1,239,698 (20.3%) (11)
Wood and timber products	898,158 (29.9%) (9)	151,562 (4.9%) (3)	1,049,720 (17.2%) (9)
Steel and other metals	444,910 (14.8%) (5)	317,531 (10.2%) (6)	762,441 (12.5%) (8)
Railroads	255,286 (8.5%) (2)	326,232 (10.5%) (4)	581,518 (9.5%) (4)
Miscellaneous corporations	227,559 (7.6%) (7)	319,162 (10.3%) (7)	546,721 (8.9%) (13)
Individuals	121,753 (4.0%) (4)	279,706 (9.0%) (8)	401,459 (6.6%) (10)
Totals	3,006,322 (100.0%) (50)	3,095,496 (99.8%) (50)	6,101,818 (99.9%) (74)

Source: Appalachian Land Ownership Study, 1980.
[a] Percentage of total surface or mineral acres held by top fifty holders.
[b] Number of holders.

Despite their profitability and power, these coal royalists are often absentee and relatively anonymous. Only one of the owners, Plateau Properties, has its headquarters in the county where most of its holdings are located—most are headquarted outside the region altogether. Only three—Pittston, Penn-Virginia, Alabama By-Products—are public companies (in the sense that they have over $1 million in assets and over 500 shareholders, and are thus required to register public information with the Securities and Exchange Commission.) Others are often family-owned, relatively small operations with merely a post office box as their address or a small office serving as their corporate headquarters. A lawyer in West Virginia describes his attempts to research the Cotiga Development Company, a Philadelphia-based operation that owns 25,081 surface acres and 39,648 mineral acres in Mingo County:

> Two years ago I wanted to do some research into the background of Cotiga.... I wanted to see the makeup of a company such as Cotiga. I went to Cotiga's office, which you have some trouble finding because it's a one-room office in a suburban home and not only is it the office for Cotiga Development Company, one of the largest landowners in Mingo County, it's also the office, according to the mailbox, for several other land companies in West Virginia. Thompson wasn't home and in talking to one of the secretaries in the office next door, she said, 'Well, he comes in one or two days a week. And sometimes there's a secretary that comes in to answer letters.' But what was interesting to me was how little it really took once you've acquired the land, to keep it going.[9]

According to interviews in Mingo County, the Cotiga holdings were acquired by an enterprising sewing-machine salesman who traveled the hills of the county early in the century trading sewing machines for land. Others of these companies also have interesting backgrounds:

Along with its affiliates Poplar Creek Coal and Winter Gap Coal Company, Coal Creek owns 64,374 acres in Anderson, Campbell, Morgan, and Scott counties, Tennessee. The company is headquartered in Knoxville and is controlled by approximately 155 shareholders throughout the United States. Most of its properties were acquired before the turn of the century, and have remained virtually the same since that time.[10] The Brimstone Company is owned primarily by John Rollins, a Delaware businessman and financier who also controls the Orkin Pest Control Company, trucking lines, Jamaican resorts, and a series of television and radio stations. Rollins acquired the 40,261 acres in Scott and Morgan counties, Tennessee, from the family of Senator Howard Baker in 1972. Senator Baker was a principal partner in the operations until 1977, when charges of conflicts of interest were raised concerning mining and potential recreation developments on the property and legislation supported by the senator.[11] Kentucky River Coal and Coke Company, located in Lexington, Kentucky, owns thousands of acres of land and mineral rights throughout eastern Kentucky—as many as 180,000 acres according to some published reports. This survey found 82,551 recorded on the tax rolls. Most of this property was obtained by John C. C. Mayo, a schoolteacher from Paintsville, Kentucky, who in the late nineteenth century received backing from eastern financiers, becoming one of east Kentucky's most successful coal buyers.[12] Kentenia Corporation, owning 25,335 acres, primarily in Harlan County, is based in Boston, Massachusetts. The company was founded in the early 1900s by Warren Delano, a wealthy Northerner and uncle of Franklin Delano Roosevelt, who invested heavily in the eastern Kentucky region.[13]

Historically, most of these coal land companies have held their land and minerals for decades, many since before the turn of the century. However, the last decade has seen in Appalachia a new wave of corporate amalgamation in the coalfields. With the energy crisis, as more corporations, often multinational ones, have moved into the energy field, a number of these coal land companies have been bought by larger interests:

In east Tennessee, the 50,940 acres of Tennessee Land and Mining, owned for decades by a family from Scarsdale, New York, has been bought by the Koppers Company, a multinational metal and chemical corporation from Pittsburgh. In 1980, Koppers also bought the 36,092 acres owned by High Top Coal Company, giving it 169,376 acres in four eastern Tennessee counties.[14] In Tennessee and Kentucky, the J. M. Huber Corporation purchased the 65,000 acres of the American Association, Ltd., a British-owned firm formerly controlled by the interests of Sir Denys Flowerdew Lowson, a former lord mayor of London. American Association had developed Middlesboro and Cumberland Gap in the 1890s.[15] The largest owner found in the study, Huber owns 227,000 acres in the survey area. In Kentucky and Virginia, the properties of Virginia Iron Coal and Coke Company have been purchased by Bates Manufacturing Company. Shortly afterward, they were acquired by American Natural Resources Corporation, a diversified energy corporation from Detroit.[16] In Tennessee, a family-held coal mining and landholding company, the Tennessee Consolidation Coal Company, has been purchased by St. Joe's Minerals. St. Joe's has also signed an agreement with Scallop Coal Corporation, a subsidiary of Royal Dutch Shell, jointly to develop its coal properties throughout the region, with much of the new production possibly to be used for export.[17] In 1979, a tentative agreement was signed for the Blue Diamond Coal Company of Knoxville to be acquired by the Standard Oil Company of Indiana (AMOCO). The deal was later dropped by Standard Oil, partly because of uncertainties surrounding some of Blue Diamond's lease holdings in eastern Kentucky.[18]

The 1970s saw growing national concern over the extent of control of the nation's energy resources by a small number of holders, particularly the oil companies. In 1963, Gulf Oil took over Pittsburgh and Midway Coal Company; in the years following, other companies followed suit. According to the Office of Technology Assessment of the United States Congress, these "horizontally integrated" companies will mine about 385-465 million tons of coal by 1986, representing almost one-half of the total domestic consumption of coal used for energy purposes.

As they acquired coal companies, oil companies also gained control over vast amounts of mineral reserves. According to the President's Coal Commission, oil and gas companies now own 41.1 percent of all privately owned coal reserves in the country, concentrated primarily in the West. Six of the top ten national coal reserve owners are partially owned by larger oil and gas companies: Continental Oil, Exxon, El Paso Natural Gas, Standard Oil of California, Occidental Petroleum (Island Creek). The largest of these, Continental Oil, owns an estimated 13.7 billion tons of coal, theoretically enough to supply the nation's needs for 15 years to come.

Of these big oil companies, Continental Oil (Consolidation Coal) and Occidental Petroleum (Island Creek Coal) are in the list of the top fifty owners in the survey area, together owning 422,320 acres of surface and mineral rights. They control thousands more acres through leasing. Altogether in the survey area, eleven oil and gas companies own approximately 1,239,698 acres of surface and mineral rights combined, an average of over 100,000 acres each.

While controlling thousands of acres of coal reserves on the one hand, the oil companies are now leasing thousands of acres of oil and gas rights on the other. According to the New York Times News Service, as much as 10 million acres have already been leased in what is called the Eastern Overthrust Belt, a geologic formation running 1,000 miles along the Appalachian Mountains from Alabama to New England.[19] Exactly who is leasing how much of this oil and gas is difficult to determine, as the rights rarely appear on the tax rolls. When the leases are recorded in county deed books, they often appear in the names of individuals serving as land agents for the oil companies. However, from other evidence, it is clear that the leasing activity extends well beyond the coalfields. Speculating about the presence of oil atop "Old Smoky," *South Magazine* reports a "land war going on for drilling rights in the Appalachian region.... Gulf, Exxon, Weaver Oil and Gas Corporation of Houston are all known to be crawling the foothills in search of landowners."[20] Already, Standard Oil of Indiana has leased 122,000 acres in just four western North Carolina counties.[21]

The oil- and gas-company presence is seen, too, in the development of new synthetic fuels plants in the region. In Wayne and Lincoln counties, West Virginia, Columbia Gas has been exploring possibilities of synthetic fuel development on its over 300,000 acres of minerals. In Catlettsburg, Kentucky, Ashland Oil has spearheaded a consortium (which includes Mobil Oil, Standard Oil of Indiana, and Conoco) that has built a pilot liquefaction plant, funded primarily by United States Department of Energy funds. The Koppers Company, already one of the largest developers of synfuels technology in the country, plans one or more plants near its Tennessee properties. Depending primarily on policies at the federal level, a number of other synthetic fuel projects are likely to develop in the region. If they do, they will have major impacts on land use, as well as air and water quality, employment, and services in the communities where they are located.

The possible development extends to oil shale, which also can be used to produce oil and natural gas. Until recently, oil-shale development has only been considered a possibility for the western states, though even there has faced major environmental opposition. Now, the Department of Energy has established an Eastern Gas Shale Project in Morgantown, West Virginia, to determine the location of Appalachian deposits. Meanwhile, the leasing has already begun. In 1979, Addington Oil Company had leased 150,000 acres of oil shale in the Knobs Belt that lies just west of the coalfield counties of eastern Kentucky that this survey examined. The company is owned by Larry Addington, one of two brothers who had been involved in strip-mining in northeastern Kentucky prior to selling out to Ashland Oil Company for a reported $13 million. Controversy over the terms of the leases led to an unprecedented order by the Kentucky Consumer Protection Division to allow landowners to cancel or renegotiate the agreements.[22]

While oil and gas companies may be scrambling for the mineral rights underground, there is also renewed interest by the timber companies in the Southeastern and Appalachian forest resources above ground. Evidence of this shift to the South is seen in the move of the headquarters of Georgia-Pacific, one of the large landholders in the survey, from Portland, Oregon, to Atlanta, Georgia. According to industry reports, other companies like Weyerhauser, Boise Cascade, Crown Zellerbach, and International Paper, are also expanding their holdings in the Southeast.

The timber companies already own substantial acreage in the region. In the eighty counties surveyed, seven companies—J. M. Huber, Bowaters, Georgia-Pacific, Gulf States, Weyerhauser, Champion International, and Mead—own 898,158 acres of surface lands and 151,562 acres of mineral rights, much of it located in southern Tennessee and northern Alabama. While using the land primarily for logging and timber growth, they may lease the minerals for mining.

Much of this corporate-owned timberland was obtained at the turn of the century, when railroads opened the vast Appalachian hardwoods to commercial exploitation. Another wave of timber-company buying occurred during the Depression. Often, as the Alabama study shows, the timber interests were able to get the land "for taxes" in court-ordered sales. When these lands were timbered out, the companies moved to the Northwest for much of their production. In many counties like Shelby County, Alabama, though, timber-company ownership has continued to dominate the development of the local economy much the same as the coal-company ownership or oil- and gas-company ownership to the north.

The new wave of timber-industry expansion into Appalachia and the South is brought on by a number of factors, including closer access to Atlantic ports and cheaper labor. Landownership patterns, however, are an important ingredient. According to the Southern Forest Institute, in the Northwest, where much of the timber is in government ownership, the RARE II study (Roadless Area Review Evaluation) and other environmental controversies are inhibiting timber production. In Appalachia, even given the large holdings by the Forest Service and the timber industry, other private owners still own a large majority of the forest lands potentially available for commercial cutting. If present trends continue, the timber companies will likely be seeking greater control, through leasing or buying, of these timber resources.

Traditionally in Central Appalachia, steel companies have joined the coal companies in the ownership of coal lands. Upon their properties, they have developed their own "captive" mines to gain the coal needed for steel processing. Often coal camps or coal communities like Jenkins, Kentucky, or Gary, West Virginia, were developed and owned by the steel companies. Five steel companies—U.S. Steel, Bethlehem, Lykes Resources, National Steel, and Republic Steel—own 342,000 acres in the eighty-county survey area.

Land and Minerals

While the steel industry does not appear to be expanding its holdings, other metals companies have been investing in the region's land and minerals, particularly since the advent of the energy crisis. The largest of these is Koppers Company, which is, as mentioned, a diversified metals and chemicals company with extensive holdings in Tennessee. Also in Tennessee, Consolidated Goldfields, a subsidiary of London-based Goldfield Mining Corporation, a company which has major investments in South African gold mining, has recently obtained 26,706 acres.

Though the main concentration of holdings by steel and metal companies is in the coal fields, there are corporate holdings of other minerals. Reynolds Metals, for instance, owns 58,000 acres in Mitchell County, North Carolina, where mica and feldspar are prevalent. More recently in the Grandfather Mountain and Spruce Pine areas of western North Carolina, a number of companies have been prospecting for uranium. According to the Department of Energy (DOE) the two areas have the potential of producing at least 14,000 tons of uranium annually.[23]

According to the President's Coal Commission, railroads are second only to the oil and gas companies in ownership of coal reserves—owning 17.4 percent of known reserves. Many of these are in the West, where lands were given to them a century ago to encourage the building of railroads. The railroads also are large owners in Appalachia, where they often joined other corporations before the turn of the century in the development of coal properties on which they themselves mined the coal needed to fire their steam locomotives.

Today the railroads in Appalachia primarily lease the coal to other energy companies, benefiting both from the royalties gained in mining and from rates charged for hauling the resource. An example may be found in the Norfolk and Western Railroad (N&W) which, through its subsidiary Pocahontas Land, owns over 280,000 acres in the counties sampled in West Virginia, Kentucky, and Virginia. In Martin County, Kentucky, "Poky" (as Pocahontas is called) owns almost 50,000 acres of surface rights and 81,000 acres of minerals—together equal to 89 percent of the surface acres in the county. The minerals are leased to subsidiaries of MAPCO Oil Company, who have recently announced plans for exporting Martin County coal, likely using N&W's rail-to-port facilities to do so. Perhaps because of the anticipated rise in the export market, N&W is reportedly obtaining new properties, such as the Kentenia Corporation in Harlan County. When the holdings of Chessie Systems (a combination of Chesapeake and Ohio Railway Corporation and Baltimore & Ohio Railway Company who operate the Western Pocahontas Corporation), Southern (now merged with N&W), and Louisville and Nashville railroads are added, four railroads in the top fifty holders own 581,518 acres of combined surface and mineral lands in the survey area.[24]

The miscellaneous category of corporations in the list of top 50 owners illustrates a diverse array of the other corporate interests with holdings in the region. They include: a chemicals corporation (Union Carbide); a utility (the Southern Company); general real estate and property developers (like

Crescent Land and Eastern Property Trading); and financial institutions (like the Boston Shamuts National Bank).

In the counties with the highest level of coal reserves, 50 percent of the land in the sample was corporately held; many of these major coal counties are located in Central Appalachia, where the corporate owners have been relatively unchanging for decades. Two trends have been identified in this study, however, that are likely to bring major changes in the corporate landownership patterns in the region.

First, with growing competition for domestic exploitation of energy and natural resources, corporate ownership and control of land and minerals is rapidly spreading from the heartland of Central Appalachia to other parts of the region. There are many examples of such corporate expansion: in the West Virginia Highlands in counties like Braxton and Randolph, Exxon and other companies have leased or obtained thousands of new acres for coal developments; in southern Virginia and western North Carolina, as has been reported, numerous companies are scrambling for control of oil and gas rights or other minerals, like uranium; on the southern Tennessee plateau, American Metal Climax, Inc., has attempted to develop the largest strip mine in Appalachia—thus far halted by citizen and state opposition. In northern Alabama, traditionally a prime agricultural valley, coal resources have been discovered, resulting in land speculation along Sand Mountain, in Dekalb County, or in the more developed areas of Marshall County. Further to the west and south, in Alabama and Mississippi, three oil companies have obtained control of millions of acres of lignite rights; while back into the Knobs of central Kentucky, several hundred thousand more acres of mineral rights have been leased by another oil company for possible oil-shale development.

Many of these areas on the "periphery" of Central Appalachia have been characterized in the past by individual ownership of land, or possibly by government ownership. The new corporate intrusion carries with it new conflicts, growing out of a struggle over how the land is to be used and to whose benefit. In many ways, the changes now occurring along the edge of Central Appalachia are similar to those undergone in the heartland of the region at the turn of the century, when ownership of land and minerals there passed into the hands of the corporations.

In these Central Appalachian counties, another important transition is occurring, with potentially significant impacts in the future. As has been seen in the discussion of the corporate owners of Appalachia, many of the traditional holders of land and minerals are being acquired by larger corporate units, chiefly the oil, gas, and energy companies. The new corporate owners bring to the region an equally new scale of capital investment, technology and corporate power. With the concentration of corporate control, single corporate decisions will by themselves be able to alter the course of an area's development more than ever before. Already such impacts can be seen in West Virginia, where Occidental Petroleum's (Island Creek's) plans for a 60,000-acre mountaintop-removal strip mine will obliterate one community and physically alter parts of Mingo and Logan counties. The far-reaching corporate power can also be witnessed in northeast Tennessee, where a decision by Koppers to build five synthetic-fuel plants on the 200,000 acres it has quietly obtained in the area over the last decade, will alter the employment, environment, and land use of the area for years to come, should the plans go through.

With the new corporate control comes another factor, important to the response of citizens or local governments. In the past, corporate decisions regarding the development of land and mineral properties have involved a relatively simple calculus of profitability, government regulations, labor supply, and community relations. Now, more global factors will be brought into play, with corporate decisions taking into consideration matters ranging from the state of Middle Eastern politics to the relative profitability of multiple corporate operations in various countries. As a consequence, the new corporate ownership brings to Appalachia greater powerlessness of citizens or local governments to influence corporate decisions, and carries with it a greater dependency of the region's people upon the power of multinationals like Koppers, Exxon, Gulf, Continental Oil, Occidental Petroleum, St. Joes Minerals, Standard Oil, Royal Dutch Shell, and others.

GOVERNMENT AND PRIVATE NONPROFIT OWNERSHIP

Despite the extent of corporate control in the region, the United States government is the single largest owner of land in Appalachia. States also own large amounts of land, in parks and wildlife areas, as do private, nonprofit institutions such as churches, universities, or the Boy Scouts. How extensive is this government and nonprofit ownership? Where is it the most prevalent?

Of the land surveyed, some 2,137,868 acres were owned by government or nonprofit groups with holdings of 20 acres or more. Of these over 2 million acres, some 97 percent are owned by only ten government agencies (listed in Table 12), making the private, nonprofit sector almost negligible.[25] Of these agencies, the United States Forest Service (USFS) is the largest single owner of land in the Appalachian Region, owning 1.2 million acres in the survey area. The United States Department of Interior owns land principally for national parks, of which the Smoky Mountain National Park is the largest. The TVA land lies primarily along the rivers and the agency's dams in the valley; while the United States Department of Energy land surrounds the top-secret nuclear processing plants in Oak Ridge, Tennessee.

Like corporate ownership, the extensiveness of public ownership varies greatly among states, particular counties, and types of counties. Government and private, nonprofit ownership is especially high in the western North Carolina mountains. Of the land sampled in twelve counties there, 40.5 percent—representing 20.3 percent of the total land—is in this category of ownership, most of it held by the USFS. Western North Carolina also tends more than any other state to attract private, nonprofit holdings

Table 12. Government Ownership of Land

Agency	Acres	States
U.S. Forest Service (U.S. Dept. of Agriculture)	1,195,113	Ala., Ky., N.C., Va., W.Va.
National Park Service (U.S. Department of Interior)	317,111	N.C., Ky., Va.
Tennessee Valley Authority	175,556	Ala., Tenn., N.C.
State of Tennessee	173,594	
U.S. Army Corps of Engineers	55,565	Ky., Va.
State of Kentucky	53,661	
U.S. Department of Energy	45,975	Tenn.
Cherokee Indian Reservation	29,405	N.C.
State of Virginia	29,030	
State of West Virginia	8,486	
Total	2,083,496	

Source: Appalachian Land Ownership Study, 1980.

such as religious groups who use the land for church camps, retreats, and recreation purposes. Though not in the survey area, the case of Buncombe County in North Carolina is instructive. According to the tax assessor, there are over 8,000 parcels of tax-exempt land held by owners who claim a religious purpose.[26] Table 13 lists the counties with the largest amount of land in government or private, nonprofit hands. Of these, Swain County, North Carolina, demonstrates the pattern most dramatically. There, 81.5 percent of the county is owned by government agencies, including the national parks and Forest Service, and land held in trust for the Cherokee Indian Reservation.

The extent of public ownership is strongly associated with certain types of counties, and negatively associated with others. One might expect, for instance, that a high degree of government ownership, especially by such agencies as the Park Service and Forest Service, would be associated with a high degree of tourism and recreation. These government lands attract those interested in hiking, fishing, hunting, and natural beauty. In turn, commerical recreation and tourist industries spring up to cater to the outside visitors, and may come to dominate the service sector of the county. The data show this association to be the case. For counties where a high proportion of the economy is based on tourism, 29 percent of the sample is publicly held. This is double the rate of government ownership in high agriculture counties, and triple the rate in the major coal counties.[27]

As in the case of corporate ownership, government ownership is expanding. The TVA and the U.S. Army Corps of Engineers seek more rivers to dam, and land to flood. The USFS continually buys land in counties where it already has large holdings, or where it plans to develop areas like

Table 13. Counties with Major Government and Private Nonprofit Ownership

	Percentage of County Surface	Percentage of Sample	Total Acres
Swain, N.C.	81.5	86.1	273,201
Clay, N.C.	47.9	73.1	64,059
Randolph, W.Va.	30.0	39.3	180,000
Smyth, Va.	30.0	45.5	83,564
Bland, Va.	29.6	41.5	70,000
Summers, W.Va.	28.3	52.2	63,380
Cleburn, Ala.	22.6	31.3	82,917
Winston, Ala.	22.5	37.1	88,577
Marshall, Ala.	22.5	52.3	82,259
Wythe, Va.	19.9	34.2	58,678

Source: Appalachian Land Ownership Study, 1980.

the Mount Rogers Recreation Area, to attract more tourists. The expansion of government ownership has been a volatile issue, especially amongst local landowners, who question who is to benefit. A western North Carolina resident commented: "Well, I tell you. I don't know if it has been very much good or not. Just to be plain with you. The farmer can't haul anything over it. It's a tourist road, and the farmers aren't allowed to go on there with a load and a funeral procession can't go on the Parkway. So, what benefit is it to the labor, commonplace people.... The Parkway has brought a lot of tourists and maybe some money.... I haven't seen none of it but I guess it has. I don't use the Parkway though. It's only for sightseers and tourists. It has added to their pleasures but as far as helping the labor class of people, it ain't worth it."[28]

Ownership by government and private, nonprofit owners also applies to mineral rights, though in many cases the extent of mineral ownership is difficult to determine. Of all of the public/nonprofit acres in the sample, for instance, only 39,243 acres of mineral rights were listed, held by thirty-nine owners. Yet, other data show that the government and private, nonprofit ownership of mineral rights is far more extensive than this, particularly under the national forests.

In the West, federal leasing policy of government-owned minerals has been a major issue. Local communities, environmentalists and others have been concerned that not enough attention is being paid in federal decision making to social and environmental impacts of mining activities.

In Appalachia, where government ownership is not as extensive as in the West, concern over federal leasing has not been as widespread, though it has been an issue in some communities. Often, companies are allowed to deep-mine coal under Forest Service land as long as entryways are driven from land owned by adjacent private owners, and as long as the federal forest is not disturbed. With a new wave of leasing in the region, pressure to exploit more of these government-owned minerals is likely to increase. Already, in southwest Virginia, over 120,000 acres of federal forest land are under consideration for oil and gas leasing, and in western North Carolina, 122,000 acres, much of it also under Forest Service land, has already been obtained by Amoco.

Controversy over mining in the national forests is also likely to rise in cases where private owners lay claim to minerals under the government lands. In several well-publicized instances, conflict has emerged as to which interest should take precedent—private owners' desire to exploit their mineral claims, or the public's claim to protection of the environment. In McCreary County, Kentucky, the Greenwood Mining Company, owned by Stearns Coal and Lumber, has fought to strip-mine coal it owns under the Daniel Boone Forest. More recently, in Scott County, Virginia, controversy has emerged over a Forest Service decision to allow a private owner claiming mineral rights under part of Devil's Fork to prospect for uranium.

The issue of private mining on public lands affects not only federal holdings. In Tennessee, representatives of a number of state agencies have been meeting regularly to set up guidelines for the leasing of minerals under state-owned lands. Environmental groups are worried that such a move will open the door for strip-mining of the coal reserves that lie under the 173,000 acres owned by the state of Tennessee along the Cumberland Plateau. In addition to government owners, several private, nonprofit owners of mineral rights were discovered in the survey. The largest of these is Harvard University, which owns 11,182 acres of oil and gas rights in Johnson and Martin Counties, in eastern Kentucky, which were left to the university by a wealthy Northeastern family.

INDIVIDUAL OWNERSHIP

The ownership of land by corporations and government leaves little for the local Appalachian. "The land companies won't let private citizens have the land at any price: a poor person can't deal with them," a retired coal miner said. Under one-half of the land in our sample is owned by individuals, and under one-half of that is owned by local individuals.

At first reading the data might suggest otherwise: over 30,000 individuals in the sample own 5,925,470 acres, or 45 percent of the land sampled. This apparently widespread individual ownership of land, however, is deceptive. The "individual" category, it should be remembered, represents holdings of two types: the local landholders of 250 acres or more, and the out-of-county owners of 20 acres or more. The vast majority of these individual owners—about 25,000 of them—are in the absentee category, owning 56 percent of the individual land in the sample. Some 90 percent of these absentee holders fall in the category of relatively small absentee owners, owning between 20 and 250 acres. This category (which was not collected for the local owners) accounts for 1,682,088 acres of 28 percent of the individual land surveyed.

Land and Minerals

A closer look at the data, then, does not necessarily support the stereotypical image of extensive individual local landholdings in the region. Only 5,079 of the 30,175 individual owners live in the counties where their holdings are located. Their holdings (above 250 acres each) total just 10 percent of the total acreage in the eighty counties. In North Carolina, only 3.4 percent of the land in the twelve counties studied is owned by these local individuals; in Alabama, the figure rises to 13.1 percent. In none of the counties do local individuals with over 250 acres account for over 30 percent of the county surface.

Just as coal lands have been associated with corporate ownership, and public lands with recreation and tourism, so we might expect this individual category to be associated with agricultural counties. On the whole, as the later chapter on agriculture shows, farming in Appalachia has not been taken over by agribusiness, as it has in some parts of the country. Also, it is where mining and federal ownership are not occurring that agriculture is still strong. (See Table 14.)

Generally, the expectations can be upheld. In the high agricultural counties, 53 percent of the land sampled is owned by individuals. This is substantially higher than in the high coal counties, where only 40 percent of the sample is individually held and slightly higher than in the high tourism counties, where 48 percent is individually owned. Similarly the degree of individual ownership in the high agricultural counties is much greater than the low agricultural counties: 53 percent compared to 38 percent. Perhaps more appropriate is to see what percentage of land in agricultural counties is held by local individuals, as the local owners are the most likely to be actually farming the land. This also shows the same pattern: in agricultural counties, 25 percent of the sample is controlled by local individuals, in tourism counties 20 percent, and in coal counties only 18 percent.

Table 14. Counties with High Percentage of Local, Individual Holdings

	Percentage of County Surface	Percentage of Sample	Total Acres
Jackson, Ala.	27.9	41.1	192,928
Tazewell, Va.	25.4	41.5	85,040
Mineral, W.Va.	24.2	47.6	51,166
Fayette, Ala.	22.2	40.6	89,112
Scott, Tenn.	19.2	25.7	66,802
Fentress, Tenn.	19.0	25.6	60,464
Jefferson, W.Va.	18.9	52.4	25,569
Bland, Va.	18.8	26.3	44,335
Cumberland, Tenn.	18.0	27.9	78,123
Lamar, Ala.	17.4	31.2	67,333
Cherokee, Ala.	17.4	26.5	61,830

Source: Appalachian Land Ownership Study, 1980.

In sum, then, the romantic image of owners living upon and working their medium-sized family holdings in Appalachia is not entirely accurate. Local individual ownership, where it does still occur, is associated with agricultural production. But these landowners of the region are under pressure: corporate ownership, often for energy and resource exploitation, and government ownership, with associated tourism and recreation developments, threaten the access people in the region have to the land and the control they exercise over its use. While only 1 percent of the local population joins corporate, government, and absentee holders to own over half the land, the other 99 percent of the population are very much affected by existing and changing ownership patterns. The nature of these effects of landownership upon rural Appalachian communities in areas of land use, property taxation and services, economic development, housing, and environment will be considered in the following chapters.

THREE

Who Bears the Tax Burden?

One of the major policy areas related to the ownership and use of land is its taxation. Historically and today, the taxation of property is the primary source of locally generated revenues for county governments, providing funding for public services such as education, roads, welfare, health, sewage. In general across the country, the proportion of the tax which actually falls on the land is small, probably less than 20 percent according to some reports.[1] Buildings and other forms of real property provide the bulk of the tax base. However, in rural areas, where improvements have not been made upon the land to the same degree as in cities, taxation of the land itself is a principal revenue source. In this survey, 50 percent of the property taxes recorded were derived from the land surface; taxes on mineral rights beneath the land accounted for 26 percent of the property taxes, and taxes on improvements only 24 percent.

Across the nation, of course, rising property taxes have provoked citizens' outcry, while at the same time lack of funds has thrown local governments into fiscal crisis. In the last twenty years, according to the 1977 Census of Governments, property values for tax purposes have increased 339 percent. From 1971 to 1976 they increased 71 percent. County taxes (about 81 percent generated from property taxes) rose 59 percent in the same period. Despite the rising local taxes, the proportion of county budgets supported by the property tax declined from 41 percent in 1966, to 36 percent in 1971, to 31 percent in 1976. "As property taxes exhibit the conflicting trends of decreasing proportion and increasing amount,"[2] local governments must either turn to federal and state sources for additional support or cut existing services.

One might not expect the fiscal crisis of local governments to be as great in Appalachia as in other parts of the country. Appalachia's mineral wealth alone offers the prospect of significant income for local governments. The owners of the wealth, as has been seen, are often large and profitable corporations, or absentee owners holding the resources for speculative value, offering the possibility of increasing taxation without overburdening already pinched small homeowners. A relatively sparse rural population may avoid some of the costly demands of urban areas.

Despite the wealth of Appalachia, however, the region's local governments remain poor. Funds are lacking for even minimal services found in other parts of the country. The reason for the disparity lies in the failure of the tax system to tax adequately and equitably the region's property wealth.

PROPERTY TAXATION OF SURFACE LANDS

Table 15 provides a short summary of the laws pertaining to property taxation in each of the survey states. According to the law in each state, land is to be appraised at fair and actual value. In Alabama and Tennessee percentage rates are set to establish what proportion of the value of various classes of property can actually be taxed. In theory, the assessment rate is to lower the burden carried by the residential and agricultural owners, while raising the burden for utilities and for commercial property. In actuality, of course, the "true and actual" value of surface lands as recorded on the tax books is low. In Tennessee, Kentucky, and Virginia, the average value

Table 15. Legal Basis for Assessed Value of Realty, by State

State	Basis
Alabama	Fair and reasonable market value. Effective in 1972, the following percentages therof apply for the types of realty indicated: Class 1, utilities used in business—30% (except in eight counties, where the level is 35%). Class 2, property not otherwise classified—25%. Class 3, agricultural, forest, and residential—15%.
Kentucky	Fair cash value.
North Carolina	True value in money.
Tennessee	Effective 1 January 1973: Percentages of actual value, as follows: Public utilities 55 percent Industrial and commercial 40 percent Farm and residential 25 percent
Virginia	Fair market value.
West Virginia	True and actual value, but four classes of property, each subject to a specified rate limit as follows, amounts per $100 of assessed value: Class 1, personalty—$.50. Class 2, owner-occupied residential property, including farms—$1.00. Class 3, all property outside municipalities, other than 1 and 2—$1.50. Class 4, all property inside municipalities, other than 1 and 2—$2.00.

Source: 1977 Census of Governments.

The Tax Burden

of an acre of land in the sample was under $100, while the going price of a piece of rural land can easily be ten to twenty times as high.

To deal with the problem of undervalued property appraisals, many counties in the region recently have undergone reappraisal by independent outside appraisal firms. Still, however, glaring examples are found of the failure of assessments to keep up with increasing values. One illustration is Martin County, Kentucky, where the Martiki Coal Company, a subsidiary of MAPCO, Inc, bought 154.25 acres in five different transactions during 1978-79. The total bill: $425,500, or $2,579 per acre. However, Martiki's entire 5,856 acres in the county are only appraised for tax purposes at $50/acre—less than 1/50 of the value of the recent transactions.

Valuation by itself, though, is a crude means of comparing property-tax structures across state and county lines. One county may have a practice of setting low values and compensating through high tax rates; other counties may assess at a value closer to actual value, while setting the tax rate at a lower level. For this reason, the more accurate way to analyze taxes in a multistate and multicounty study is to look at the "bottom line": the actual taxes paid per acre of land. In so doing, some clear patterns emerge about surface taxation of rural land in the eighty counties studied.

In general, the taxes paid on rural lands are relatively low. Almost a quarter of the owners in the study pay less than $.25 per acre for their land; only a little more than one-third pay over $1.00 per acre. Overall, the amount of taxes paid per acre is only $.90 per acre for the taxable land in the study. In Alabama, the average tax per acre is only $.49 (before the recent reassessment). In North Carolina it rises to $2.07. In other states the average per surface acre is as follows: Kentucky, $.79; Tennessee, $.79; Virginia, $.84; and West Virginia, $1.28. (See Table 16.)

Overall, corporations pay more per acre than do individuals, $1.03 per acre compared to $.78. However, there is not a consistent pattern. In Virginia, for instance, corporations pay only $.67 per acre, while individuals pay $.94; and in Tennessee, corporations pay $.68 per acre, while individuals pay $.89 per acre.

When residence of the owners is considered, one finds that in four of six states, out-of-state owners pay less per acre than do local owners of land in the sample. In Alabama, local owners pay $.64 per acre, while out-of-state owners pay only about 60 percent of that—$.39 per acre. In Virginia, there also is found a large discrepancy: local owners pay $1.04 per acre, while out-of-state owners pay only $.66. Similar patterns are found in North Carolina and Tennessee. Only in Kentucky and West Virginia do the absentee owners pay more per acre than do the local owners (and in West Virginia it may be due to the fact that coal appraisals in that state are sometimes reflected in the surface values).

When residence is considered, one also gets a different perspective on the taxes corporations pay: on the whole, out-of-state corporations—many of whom are holding the land for its speculative and mineral value—pay far less per acre than do local corporations, many of whom may be using the land for industry; or than local individuals, many of whom are using

Table 16. Property Taxes Paid Per Acre of Surface Land by State and by Type and Residence of Owner (in dollars)

	In-County	Out-of-County In-State	Out-of-State	All
		Individuals		
Alabama	.46	.42	.35	.43
Kentucky	.63	.78	.59	.66
North Carolina	1.53	2.01	1.81	1.84
Tennessee	.96	.87	.79	.89
Virginia	1.02	.86	.85	.94
West Virginia	.51	.72	.71	.56
All individuals	.72	.82	.84	.78
		Corporations		
Alabama	1.40	.43	.42	.59
Kentucky	1.10	.59	.97	.92
North Carolina	3.26	3.18	1.82	2.61
Tennessee	.96	.67	.62	.68
Virginia	1.12	.83	.53	.67
West Virginia	1.30	1.88	1.60	1.59
All corporations	1.37	1.06	.94	1.03
		All Taxable Surface (Individual + Corporate)		
Alabama	.64	.42	.39	.49
Kentucky	.69	.72	.86	.79
North Carolina	2.10	2.38	1.82	2.07
Tennessee	.96	.81	.66	.79
Virginia	1.04	.85	.66	.84
West Virginia	.84	1.61	1.51	1.28
Total sample	.87	.92	.90	.90

Source: Appalachian Land Ownership Study, 1980.

the land for housing. In Alabama, for instance, out-of-state corporations pay only $.42 per acre—less than one-third that paid by local corporations, and slightly less than the rate paid by local individuals. In Virginia, absentee corporations pay $.53 per acre for their land, while local individuals and local corporations pay about twice that, or $1.02 and $1.12, respectively.

Not only do absentee owners pay less than local owners (with out-of-state corporations often paying least of all), but another related pattern is also found: larger owners tend to pay less per acre than do the smaller owners. As Table 17 shows, 34 percent of the owners with over 1,000 acres each pay under $.25 per acre in taxes, while only 23 percent pay more than $1.00 per acre. For the smaller owners with under 250 acres each the reverse

The Tax Burden

Table 17. Taxes Paid Per Surface Acre, by Size of Owner's Holdings

Holdings	$.25 or less/Acre	$.26–.50/Acre	$.51–1.00/Acre	Over $1.00/Acre	Total
250 Acres or less	5,052[a] (65.2%)[b] (20.4%)[c]	5,061 (78.4%) (20.4%)	5,635 (73.1%) (22.8%)	9,013 (80.1%) (36.4%)	24,761 (74.6%)[d]
251–500 Acres	1,592 (20.5%) (30.6%)	890 (13.8%) (17.1%)	1,232 (16.1%) (23.7%)	1,480 (13.1%) (28.5%)	5,194 (15.7%)
501–1,000 Acres	637 (8.2%) (34.2%)	297 (4.6%) (15.9%)	474 (6.1%) (25.5%)	545 (4.0%) (24.4%)	1,862 (5.6%)
Over 1,000 Acres	466 (6.0%) (34.4%)	203 (4.6%) (15.0%)	366 (4.7%) (27.1%)	314 (2.8%) (23.4%)	1,351 (4.1%)
Total	7,747 (23.4%)[d]	6,451 (19.4%)	7.707 (23.2%)	11,263 (34.0%)	33,168 (100.0%)

Chi Square = 628, Probability = .0001
Source: Appalachian Land Ownership Study, 1980.
[a] Number of Owners.
[b] Percentage of owners in tax rate category (column).
[c] Percentage of owners in holdings size category (row).
[d] Percentage of total owners.

pattern is true: only 20 percent pay under $.25 per acre, while 36 percent pay more than $1.00 per acre. This pattern—the bigger the owner the less the taxes—holds particularly true for the Tennessee counties in the survey. There, of owners with more than 1,000 acres, 23 percent pay over $1.00 per acre, as in the overall sample, but of the small owners with 250 acres or less, 52 percent pay more than $1.00 per acre of surface owned.

Why the discrepancy? Why do the absentee and the large owners tend to pay less per acre of surface land than the more local smaller owners? There are many reasons, of course, but part of the answer lies in the use to which the land is put.

The primary means by which rural assessors determine value is through recent sales on the market. Value is fixed according to what willing buyers would pay willing sellers in arms-length transactions. However, this presents a problem in assessing the value of the vast tracts of land held primarily by absentee corporations in many parts of Appalachia: large tracts of land may be traded rarely. Interviews in the case studies show time and again that the large owners have held the land for decades and do not want to sell. The assessments on the land reflect past values for rural property, when land was abundant and relatively cheap, not the values of today—where land is becoming increasingly in demand and more valuable. At least some assessors have ruled that only one or two transfers do not determine a pattern, and they have refused to consider certain recent sales in making their assessments, despite the fact that alone one transfer of these vast tracts of land can affect large portions of a county. While the market yardstick is used to value land, in some areas the concentrated control of land in a few unchanging hands has, in effect, taken the land out of the market, thus rendering the yardstick ineffective. As a result, not only do larger tracts go underassessed, but competition increases for the land that *is* being bought and sold, driving its values higher and higher.

The assessed value of the large absentee tracts remains low for another reason—on the whole, these tracts are being held for speculative value, or for the value of the minerals underneath (which is also underassessed). The owners do little to improve the value of the land—it is classed simply as woodland or mountain land, receiving a low appraised value, and taxed at an average of only $.68 per acre. On the other hand, local owners tend to improve the land with homes and other buildings, having the effect of increasing its value. For individually owned land, local owners tend to build on their land, and to make more valuable improvements, thus raising their property assessments.

Even though the local land in the survey was only the plots 250 acres or above, 92 percent of the locally owned plots have building improvements on them, with an average tax of $101.06. On the other hand, only 33 percent of the parcels owned by out-of-state individuals have buildings, taxed at a rate of only $39.16 each. The pattern adds to the already regressive nature of the property tax: local residential owners who have less land pay more for it—an average of $1.16 per acre according to the survey.

It is partly to overcome this regressive nature of the tax that various states have adopted classification systems whereby land is assessed at different percentages of its value according to its use. In Tennessee for instance, commercial and industrial land is to be assessed at 40 percent of its value, while residential land and agricultural land is assessed at only 25 percent. Alabama has a similar classification system, and in Kentucky, an agricultural use provision is meant to give special breaks to agricultural land. While the principle of classification according to use is an accepted one, its misuse in Appalachia has increased rather than eased the property-tax inequities:

In Tennessee, vast tracts of land owned for mineral development by coal land companies and energy producers have been routinely assessed as "farmland," paying at a 25 percent rate rather than the 40 percent assessment rate required for industrial and commercial produces. A citizen's complaint in 1978 resulted in a state ruling that commercial rates should be applied when the land is leased for mining purposes. However, the decision may not lead to change: local assessors have been slow to implement the rule, and may lack reliable information as to which lands are actually leased for mining.

The Tax Burden

In Kentucky, the legislature in 1968 passed an amendment to the Kentucky Constitution, section 172A, which allowed assessments at less than full cash value for land used for agricultural or horticultural purposes. The purpose of the amendment was to lessen the impact of property taxes on the farmer. By statute, only corporations organized primarily for agricultural purposes and which derive a substantial portion of their income from farming or horticulture may benefit from these reduced taxes. In practice, however, east Kentucky assessors have applied the provision to any owners of more than five to fifteen acres (depending on the county). The major beneficiaries of the practice, of course, are the energy giants and coal landholders, who practice no agriculture. Since 1968, in eastern Kentucky, these large coal- and landowners have received up to 50 percent reduction in property taxes due to this provision.

In Alabama, similar current use provisions are at work. Speculatively held timber- and mineral lands are given the low assessment rate designed to protect forest areas. As a result, the land is assessed at $22.70 an acre, and yields only $.59 per acre in taxes.

The ultimate effect of this pattern can be seen in the Table 18 which gives the taxes per acre of surface land by its use, as defined on the tax rolls.[3] While mineral land under development and commercial/industrial land are taxed at a higher rate than woodland or farmland, relatively few acres— 33 percent—are classified in that category. The largest portion—58 percent —of the land is in the woodland and agriculture category, despite the fact that the principal owners of the land are holding it for energy purposes, or for speculation, not for agriculture at all.

If the larger, absentee owners are the beneficiaries of surface taxation patterns in Appalachia, they also fight to keep it that way. During the course of this particular study, the tax issue was perhaps the most controversial in Alabama, where in 1978 the legislature passed Amendment 373, a "Tax Relief Package" that had the effect of placing a "lid" on the amount values could be increased through a court-ordered statewide reappraisal program. The amendment was supported by a "grassroots organization"

Table 18. Taxes Per Surface Acre, by Land Use (*highest to lowest*)

Land Use	Surface Taxes/Acre	Number of Acres	Percentage of Total Classified Land
Mineral land under development	$1.97	680,344	12
Commercial/industrial	1.45	1,225,651	21
Residential	1.16	516,883	9
Woodland/forest	.68	2,350,458	40
Agricultural	.68	1,051,371	18

Source: Appalachian Land Ownership Study, 1980.

called the "Alabamians for Tax Relief Committee." Handsomely financed with a budget of $100,000, the group received much of its funding from the Farm Bureau, and from Alabama's large corporate landholders: the Gulf States Paper Company donated $3,650; Weyerhauser Company donated $1,800; International Paper gave $5,000; and Champion International gave $1,900. Though Alabama has the lowest property-tax base in the country, a before and after study of the reappraisal program shows that as a result of the Tax Relief Package the large landholders still pay little for their land. In fact, by conservative estimates, Amendment 373 provided tax relief of at least $1 million a year to the twenty-six largest landowners in the state.

Concerning the taxation of surface lands in Appalachia, then, a clear pattern emerges. Large and absentee owners pay less per acre of land than the small and local owners pay. While the reasons for the pattern may be numerous, several have been discussed: the relatively unchanging monopoly of large tracts, rendering the market approach to valuation ineffective; the failure of the large and absentee owners to improve their properties; the "misuse" of the use principle; and the organized political pressure of the large and corporate owners to keep their taxes low.

PROPERTY TAXATION OF MINERALS

If there is any place in the country, though, where one might *not* expect a property-tax crisis, it might be resource-rich Appalachia. Among other resources, the region contains massive reserves of coal, the "black gold" of the energy area. Oil and gas deposits also stretch under a number of its counties. With the nation turning more and more to domestic energy sources, the region's resources have increasing value to the nation and to the world. But despite the rapidly escalating values, Appalachia's mineral wealth remains relatively—even startlingly—property tax-free. The figures gained in this study speak for themselves:

Over 75 percent of the 3,950 owners of mineral rights in the survey pay under $.25 per mineral acre in property taxes. Some 86% pay less than $1.00 per acre. In the twelve counties in eastern Kentucky—which include some of the major coal-producing counties in the region—the average tax per acre of minerals is $.002. The total property tax on minerals for these major coal counties is a meager $1,500.

Altogether the eighty counties in the survey receive only $5.1 million in property taxes from their enormous mineral wealth (mostly from coal). Some 97 percent of this revenue comes from the thirty-seven counties classified in this study as high coal-reserve counties (i.e., counties with over 100 million tons of reserves). Twenty-two of these counties are known to have over a billion tons of coal reserves. By conservative calculations, then, the average tax per ton of known coal in the ground in these major coal counties is only $.0002 per ton.

What accounts for this situation in which Appalachia's most valuable resource, its mineral wealth, is taxed so low? Unlike surface taxation, in

The Tax Burden

which patterns could be found across the six states, the case of mineral taxation requires state-by-state examination.

In Alabama the average tax per recorded mineral acre is only $.04. Even that figure is deceptive, for it only includes minerals that have been severed from the surface ownership. Minerals owned "fee simple" with the surface are not valued at all, despite the Alabama Code, which states that "real and personal property shall be estimated at its fair and reasonable market value—taking into consideration all elements or factors bearing on such value."[4] Even the severed minerals are not taxed very highly. Usually the value of mineral rights is self-declared by the owner. Most mineral acres are valued at only $10 to $15 per acre, far less than its market value today. Moreover, most of these mineral acres are assessed at only 10 percent of the fair market value, a rate specified for agricultural, residential, and timber land according to calculations for this study. If the mineral rights in fifteen northern Alabama counties were appraised at just $100 per acre, the taxes per acre would still be only $.62, but over $50,000 a year of additional revenues would be generated.

If property taxes on minerals are low in Alabama, they are next to nothing in Kentucky—the leading coal producer in the country. In Kentucky, a 1978 state law established a uniform rate of one-tenth of one cent per $100 value on all unmined coal. The result virtually eliminates property taxation on coal in the ground: for instance, in Martin County, Kentucky, the largest coal producing county in the state, Norfolk and Western Railroad (Pocahontas Kentucky) owns 81,333 acres, equivalent to 55 percent of the county's surface. The coal is valued handsomely: $7,604,963, but the actual tax generated is only $76.05.

The 1978 legislation establishing the rate of taxation on coal reserves of one-tenth of one cent per $100 value came after failure by the state to develop a mineral taxation program. In 1976, the legislature had enacted a property tax on unmined coal of 31.5¢ per $100 value, to be administered by the state. Even at this low rate, the program was marred: only four inspectors were hired to assess the state's reserves. Unused to any taxes at all, the companies refused to cooperate: the *Courier Journal* reported on 1 June 1977 that of 7,000 tax report forms mailed to known coal owners and mining companies, only one-third or less were returned. Of those, less than 10 percent contained "adequate" responses. In 1978, the state gave up the program, turning coal valuation back to the local assessors. However, the "one-tenth of a cent" flat rate set by the legislature has effectively left the local assessors unable to generate revenue from east Kentucky's vast coal property. The situation goes on, despite the fact that east Kentucky counties are heavily subsidized by state and federal funds for even minimal services, and desperately need new property tax revenues.

In North Carolina, the average tax placed on minerals is $.12 per acre. However, there are only a few instances of recorded mineral rights: only fifteen owners controlling 207,330 acres were found in the survey. The low number of mineral acres compared with other states is because North Carolina has no coal reserves. With the current exploration in the western part of the state for other minerals—uranium, oil, gas—mineral taxation may become a more important policy issue.

Although Tennessee statutes state that minerals must be taxed as real property, this simply was not done until 1971, when a complaint by a group of east Tennessee citizens resulted in a decision by the State Board of Equalization to tax coal reserves. After the ruling, a procedure was adopted using the Hoskold formula to compute the present value of the coal underground based on the projected income stream it would bring to the owner. State staff (primarily one geologist) was delegated to help local county assessors to obtain coal reserve information and to map coal ownership.

However, according to data obtained in this study, nine years following the state's ruling most of the mineral resources still go relatively tax-free. The lack of implementation of the state's ruling has been widespread. The state staff of one person mapped only three counties before being transferred to another task; in eleven of the sixty-four counties surveyed, the full market value is still set at less than $30 per acre; in seven of the counties it is below $10 per acre. The average tax paid per mineral acre is still only $.15.

Despite the lack of implementation, important precedents and procedures have been set in Tennessee for coal taxation. Primarily as a result of citizens' pressure, taxes have been raised on some plots; coal-company equipment has been entered on the books; and the 40 percent commercial assessment rate has been applied to coal-company land leased for mining, replacing the 25 percent farmland rate there previously. If the state were to continue its program of assistance to counties, more revenues clearly would be generated.

The average taxes per acre of minerals on the taxbooks in Virginia double the average rate of any other state in the survey. However, the higher rate is deceptive, for in Virginia there is a crucial distinction between minerals under development and minerals not under development. For minerals under development (i.e., being mined), the State Department of Taxation has established procedures that give taxes ranging from $10 to $76 per acre, depending on the county. However, this is applied to under 1 percent of the mineral acres found in the survey.

No procedures have been established by the state for mineral reserves (i.e., minerals not under development). Using their own rule-of-thumb procedures, assessors have established mineral taxes ranging from $1.09 to $1.95 per acre on undeveloped minerals in the southwest Virginia coal-producing counties. While what *is* on the books may be higher per acre than other states in the survey, there are hundreds of thousands more mineral acres not recorded at all, and no mapping program has been established by the state to help local assessors determine where these mineral reserves are. The result of the failure to assess mineral reserves adequately is an enormous loss of revenue for southwest Virginia counties. Conservative estimates using formulas described below suggest that the major coal-producing counties would realize $2.4 million additional tax dollars annually if coal reserves were properly taxed.[5]

The only state in the survey area in which the problem of undertaxation of mineral reserves has received concerted attention by state government is West Virginia. There, the State Tax Department has adopted the following position: "Nature has endowed West Virginia with abundant mineral resources; coal in particular.... However, the coal industry's support of local government and schools, through property taxes, has not been realistic given the extent of the industry's mineral and fee property holdings. These huge fee and mineral properties and their assessments are a primary concern in West Virginia as an equalization problem."[6]

The first problem for the state was to determine who owned the coal reserves, and to map their location. Historically, assessors in the region had accepted the adage "you can't assess what you can't see." The state took a different position: "The problem has been that no one really was sure how to value coal in the ground since it was not generally visible and the extent and amount of coal property contained was difficult to determine. The industry always advanced the argument that it is impossible to assess property if you are not sure of that property's existence, location or volume. One of the first objects of the West Virginia Coal Appraisal and Assessment Program was to attempt to defeat the industry's arguments."[7] In 1970 the state began a program to map the ownership of mineral parcels. Then the following formula was adopted to value the coal reserves: Value of coal per acre = per ton value X (seam thickness $\times$ 1,500 tons). Per-ton value is computed based on a range of factors: British Thermal Unit (BTU) content, royalty rate, seam thickness, and so on. By the summer of 1980, thirty of the forty-four coal-bearing counties had received their reappraisal figures, resulting in approximately $8,400,000 per year accruing to the counties.

While West Virginia's coal appraisal program is unprecedented in the Appalachian region, it has been criticized on a number of counts for still providing overly conservative estimates of coal values.[8] The program has proceeded slowly, with no mandate that the counties must abide by the figures. Assessors typically put their coal on the books at 50 percent of the state's appraisals. Groups like West Virginians for Fair and Equitable Assessment of Taxes have also questioned the accuracy of the program, when the highest valuations ($756 for Harrison County) and the lowest valuations ($67 in Dodderidge County) are in contiguous counties. Despite the shortcomings, the West Virginia program shows that coal reserves can be taxed, with adequate effort. The state now can claim, "Valuation of coal properties in the completed counties more nearly reflect the real world than do valuations previously shown."[9]

MINERAL TAXATION: THE ALTERNATIVE

It is clear that there is a pattern of underassessment throughout the region, particularly in Alabama and Kentucky, though also in Virginia and Tennessee. Only West Virginia as a state has made a concerted effort to value coal in place, and its program is recent.

Some policy analysts argue the severance tax based on the number of tons produced is a more appropriate tax on coal than is the ad valorem or property tax. Tennessee, Kentucky, and Virginia each have a form of severance tax, though the procedures used and revenues generated vary immensely. The severance tax does serve to generate needed revenue, but it may not necessarily serve the same purpose as the property tax. The severance tax is placed on the producer of the coal, leaving the owners of the vast coal reserves, who lease the reserves to be mined, affected only nominally or indirectly. Moreover, in the Appalachian region, the producers are often relatively small, local operators who bear an additional tax burden, while the large, absentee coal owners from whom they lease the coal pay next to nothing to the local government. Also, from the local government's perspective, a severance tax makes the tax revenues highly dependent upon the ups and downs of the coal market. Taxation of the coal reserves in the ground, on the other hand, could provide a steady stream of revenue for years to come.

From a policy perspective, there is no question that coal in place has value—particularly in these days of high energy demand and a national program aimed at increased use of coal reserves. In a United States Bureau of the Mines booklet, Donald Colby and David Brooks write, "Generally speaking, any mineral deposit that can be exploited at a profit today, or that will become exploitable within the next few decades, has economic value.... The fact that minerals do exist for purchase, and the sale of mineral deposits and the rights to explore them proves that some economic value inheres in the resource itself."[10] However, as the figures above reflect, while the value of coal has increased rapidly in the last decade, the ad valorem taxes on the whole have not kept up: in Alabama, mineral taxes have not altered since the 1930s; in Virginia, tables used were established over ten years ago; and in Kentucky, coal taxation has regressed to the current situation.

If minerals were to be appraised, how would it be done? As are other property taxes, the ad valorem taxation of minerals is based upon the concept of the "fair market value": what a willing buyer would pay to a willing seller in a competitive market. In general, there are three accepted approaches for making such a valuation. The *cost approach* ascertains the building cost of improvements. It is only applicable to determining the value of mining operations on developed mines, and does not reflect the value of the coal in place. The *market approach* uses recent sales of comparable property to determine value. While this approach is relatively simple and is the one most often used for other property, it is usually ineffective in Appalachia, where much of the coal property has been owned by the same owners for decades, with few recent transactions. Where transactions have been made, they may not have been "arms-length"; the terms may be difficult to determine; or different geologic conditions of the coal may make them not comparable to other coal lands. A study by the West Virginia Tax Department has made the same point: "After more research in coal property sales, it was concluded that because of the limited number of sales and the difficulty of finding similar and comparable coal land sales in some counties, this concept could not be utilized in most situations."[11] The third approach, the income approach, is based on the capacity of the property to produce an income stream to the owner over a period of time. This approach is most applicable to mineral valuation.

In applying the *income approach* to mineral valuation, essentially two steps are required: (1) determining the future income of the owner, taking into account the amount of recoverable minerals, an estimated market price, and expenses to be incurred in developing the minerals, and (2) reducing the income to present worth, that is, determining what a prospective buyer would be willing to pay today for the prospective income in the future. Each of these steps may be elaborated:

When applied to the operator of a mine, determining the future income can be a complex process, involving estimating operating costs, depletion, depreciation, and so on.[12] However, when applied to the owner of the resource, the process is simpler: roughly, the revenue stream is equal to the royalties received over the economic life of the coal. Thus, if an owner receives $2.00 a ton for five years, and one ton is mined yearly, the future income is $10.00. Few operating expenses or other factors are involved.

Determining the present worth of future income involves "discounting" the future income to its present value. It is the reverse of compounding principal by a given interest rate. Using the previous example, this process would determine how much $10 accrued over five years is worth today at going interest. The discount formula may also take into account factors of risk or speculation. At a 20 percent speculative interest rate, the present value of $10 accrued over five years would be $5.98.

Simply put, then, the value of coal in the ground is equal to the total royalty it will produce to the owner over time discounted back to present value. Using this approach, it is possible to estimate the current value of a coal property that hypothetically produces one ton a year. Then, applying the figure to the eighty counties in the survey, an estimate can be made of the total tax value today of coal in place in the counties studied.

In making the calculations, various assumptions must be made. These assumptions are conservative, that is, they will provide a conservative estimate of the real value of the coal in place:

1. Using predictions by the President's Coal Commission, national production can be expected to increase 28 percent by 1985, and 97 percent by 1990. From the year 2000 on, triple the rates of today's production can be expected.[13] Thus, for our hypothetical example, we can project that for every one ton mined in 1980-85, 1.28 tons will be mined in 1986–90; 1.98 tons from 1991-2000; and 3 tons from 2000.
2. When the method is applied to a specific parcel of coal, the amount of reserves present must be determined, in order to determine the estimated life of the income stream. However, on an aggregate level, the problem is less difficult: Appalachia's coal reserves are expected to last for another 200 years. For the pur-

poses of the calculations, we shall only use the income stream for the next fifty years.
3. Royalty rates to coal owners in Appalachia have increased dramatically over the last few years, reflecting the growing value of the resource. A royalty rate of $2.00 per ton is used here. To be conservative, no increase in royalty rates is projected.
4. One of the most difficult problems is to ascertain the appropriate interest or discount rate to use. The higher the interest rate, the less the present value of the future income. In order to be conservative, i.e., to err on the side of undervaluation, a discount rate of 20 percent is used, approximately 12 percent reflecting current interest rates and 8 percent to take into account unforeseen risks. Based on the 20 percent rates, discount ratios are determined from standard mathematical tables.

Using these assumptions, we may return to the hypothetical example.[14] With the assumed increasing rates of production, a parcel producing one ton of coal a year now will produce 91.6 tons over the next fifty years. At a royalty of $2.00 per ton, the total income to the owner will be $183.20. Discounted back to present value at a rate of 20% annually, the current value of the $183.20 is only $12.50. (In other words, at an interest rate of 20% compounded annually, $12.50 today will be worth $183.20 in fifty years; see Table 19).

By this method, we can estimate the present value of coal reserves in the eighty counties surveyed. Based on 1977 production levels, the eighty counties produce 195 million tons a year. At production rates predicted by the President's Commission on Coal, and the assumptions given above, the present value of the coal reserves to be mined over the next fifty years is $2.4 billion. Using current average assessment and tax rates (calculated from the sample for each state), the total property tax to be produced annually from this coal value would be $21.7 million.

Currently, property taxes from all mineral property taxes (not just coal) in the eighty counties equals only $5.1 million. Thus application of even this conservative method of calculation would more than quadruple the mineral taxes generated from the fifty-six coal-producing counties in the study. The new tax revenues would equal $16.5 million annually, or almost $300,000 per county. Eight million dollars of the new revenues would be generated in eastern Kentucky, where they are desperately needed.

If less conservative assumptions were made, the amount of revenue generated from an adequate coal appraisal program would escalate rapidly. For instance, if assessments were made on developed mines as well as the undeveloped reserves, as some attempt is made to do in Virginia and Tennessee, the amount would increase substantially. If all reserves were considered rather than just those to be mined in fifty years, or if a lower discount rate were used, the possibility of generating $50 million a year of coal property taxes in the counties studied would not be unreasonable. This would be a significant income source, equal to almost 50 percent of the total property taxes collected in these counties for 1976-77.

The Tax Burden

Table 19. Current Value of Income Stream on One Ton of Coal Per Year, Increasing over Fifty Years

Life of Income	Annual Production Rate (tons)		Royalty Rate/Ton		Discount Rate		Income
1st–5th years (1980–85)	1.00	×	2.00	×	2.99	=	$ 5.98
6th–10th years (1986–90)	1.28	×	2.00	×	1.20	=	3.07
11th–20th years (1991–2000)	1.97	×	2.00	×	.67	=	2.67
21st–30th years (2001–2010)	3.00	×	2.00	×	.109	=	.65
31st–40th years (2011–20)	3.00	×	2.00	×	.018	=	.11
41st–50th years (2021–30)	3.00	×	2.00	×	.003	=	.02
Total							$12.50

If ad valorem mineral taxation represents such a potential revenue source, why has it not been tapped? As in the case of explaining patterns of surface taxation, there is no single answer.

Partly, one suspects, the nonpayment of mineral taxes is the holdover of a historical period when the coal in the ground did not have the value that it has today. To update the assessments is a massive and complex task, requiring far more precise information than necessary for the above estimates. Local assessors simply lack the resources, the data, the staff, or the skills to do the job.

It would be wrong, however, to give the impression that the problem is merely technical. In the coal counties of the region, the coal owners traditionally have had their own way, often using their political muscle to make or break the political fortunes of local officials, especially tax assessors. In many cases, the companies have supplied assessors with their own assessments of property values, and assessors have had little choice but to accept them. Where attempts are made to alter the traditional patterns of underassessment, the coal owners may simply refuse to cooperate, as was seen in eastern Kentucky where they failed even to return tax forms regarding their properties. In cases where local assessors have pressed the matter further, they have often found themselves beaten down in appeals procedures by a battery of technical experts and lawyers far greater than what the local assessors can muster by themselves.

Where changes have been made, they have been as a result of citizens' pressure combined with state intervention. Thus far, however, these cases in most states have been isolated and inadequate. For successful action upon the problem, state and federal assistance will be needed, to provide the resources for mapping and assessing coal reserves, as well as to provide the political muscle necessary for the task. While the task may at first appear to be large and expensive, the long-term return of additional revenues to local governments could be substantial enough both to improve local services and to decrease the federal and state subsidies currently going to these counties.

THE PROBLEM OF TAX-EXEMPT LANDS

Like concentrated ownership of surface or mineral lands by private owners, a concentrated presence of tax-exempt government or private, nonprofit lands may also have negative effects upon a rural tax base. In a report on property taxation the Council of State Governments summarizes the issue: "Whether federal or state owned, exempt real property presents problems to local jurisdictions in which the property is located. Primarily, these problems are tax revenue loss, restraint of community development, and local government financial impoverishment."[15]

As indicated in the previous chapter, this study identified about 2.1 million acres of land held by government owners or by private nonprofit owners, such as churches, universities, or civic groups. The overwhelming portion of this land is government owned, usually by federal or state agencies. Of these, the largest owner—and the largest owner in the study—is the U.S. Forest Service with 1.2 million acres. While these lands are legally tax exempt (based on the landmark decision of McCullough v. Maryland), the Forest Service has accepted an obligation to make payments in lieu of taxes since the Weeks Act of 1911, which authorized the agency to share with counties revenues derived from sale of timber and other uses of its land. In 1976, Congress further enacted the Payments in Lieu of Taxes Act which, in essence, sought to guarantee that counties with Forest Service or other federal lands received a minimum of $.75 per acre of federal land in lieu of tax payments.

In the Virginia counties surveyed, the $.75 per acre of federally owned land is less than what the ad valorem tax would be if the land were privately owned. For example, if the 70,000 acres owned by the Forest Service in Bland County were taxed at the same rate as the land owned by individuals, the county would receive $.95 per acre; if the same rate were used as for out-of-state corporations, it would receive $1.06. A similar pattern is found in North Carolina. In Clay and Swain counties in North Carolina, the two counties with the highest level of federal ownership, the $.75 per acre does not compare with the $1.05 per acre tax that out-of-state corporate owners average paying or the $1.22 that out-of-state private owners average paying. If the federal agencies paid the lower rate, $98,182 additional revenues would be generated; if they paid the higher rate, the additional revenue would be $158,518.

Not only are the federal acres taxed less, but the federal ownership in turn limits the amount of land and developments that are taxable: In Clay

The Tax Burden

and Swain counties, only eight local owners own more than 250 acres each. One official in Swain County makes the point: "Eighteen percent of the county is all that's taxable. Well, we just make do. To give you an example, this year's budget requests were cut drastically because we just don't have the ability to give services I think we should." The effects of federal ownership may also be felt strongly where the Forest Service is still purchasing land, thus removing it from the tax base virtually overnight. In Wythe County, Virginia, where federal purchasing continues, the amount of revenue the county receives per acre drops from $1.22 to $.75 for every acre of forest land purchased. Members of the Mount Rogers Planning District have gone on record opposing further land acquisitions by the Forest Service until the discrepancies have been reduced.

The problem does not stop with federal lands. Counties usually receive no compensation at all for state lands within their borders. Of the six states surveyed, only North Carolina has a program of compensating local counties for state-owned land. The lack of revenue may be especially significant in places like Morgan County, Tennessee, where the state owns over 50,000 acres of land in state forest and for the maximum-security prison, yet the county receives no compensation.

While the problem is significant, its solution is often out of the reach of local citizens or officials, who feel powerless to influence congressionally established payment systems. Though Virginia has passed legislation that allows local governments the option of imposing service charges on certain exempt properties, this study found no cases where the charges had actually been made. Certainly, while other states or counties might investigate similar options, real change is not likely without federal action.

THE IMPACT OF TAX PATTERNS

Taken together, the underassessment of surface lands, failure adequately to tax minerals, and the revenue loss from concentrated federal holdings has a marked impact on local governments in Appalachia. The effect, essentially, is to produce a situation in which small owners carry a disproportionate share of the tax burden; counties turn increasingly to federal and state funds to provide revenues, while the large corporate and absentee owners of Appalachia's resources go relatively tax-free; and citizens face a poverty of needed services despite the fact that they sit upon taxable property wealth, especially in the form of coal and other natural resources.

On the whole, the data from the sample of 33,000 owners in eighty Appalachian counties substantiate this pattern: large owners contribute less to the tax base relative to what they own than do the smaller owners. Several factors, as has been seen, affect the pattern: the larger owners of land have their surface lands taxed at a lower rate per acre than the smaller owners; the larger owners tend to own the bulk of the mineral wealth, which is not adequately appraised, and tend not to develop improvements on their land. On the other hand, the smaller owners have their land taxed at a higher rate than the large owners; they are also likely to improve their land and thus to increase their taxes as well. Federal holdings, which tend to be large, pay in lieu of taxes, but at a lower rate than privately held land. The additive result is an overwhelmingly regressive property-tax system in rural Appalachia.

To help illustrate the point, the property-tax burden can be measured by dividing the percentage of taxes paid by owners in the sample by the percentage of land owned to obtain a "tax-burden ratio." As Table 20 shows, for the larger owners this ratio is low; as the landholders get smaller, the proportion of taxes paid relative to the amount of land owned increases. For instance, the top 1 percent of the owners own 22 percent of the land in the eighty counties but pay only 4.7 percent of the property taxes. The ratio of taxes paid to land owned is .21. By contrast, the bottom 1% of owners in the sample own .02 percent of the land in the survey areas but paid .23 percent of the taxes, a tax-burden ratio of 11.5. The top 5 percent of the owners owned 31.3 percent of the land and paid 7.1 percent of the taxes, for a tax-burden ratio of .23. The bottom 5 percent, owning .13 percent of the land and paying .30 percent of the property taxes, had a tax-burden ratio of ten times as high, 2.3, and so on. In general, the higher up the ownership ladder, the lower the property-tax burden relative to the amount of land owned.

Altogether, the owners in the sample (who themselves represent the larger property owners compared with the small owners not in the sample) own 53 percent of the total land surface in the eighty counties studied, yet account for only 13 percent of the total property taxes collected.

One could respond to these apparent inequities with the argument that the smaller owners are probably more likely to have improvements on their land, and thus property values contribute more to the tax base. However, the response itself helps to make the basic point: the net effect of the

Table 20. Land Owned and Property Taxes Paid, by Owners

Owners in Sample	Percentage Surface Land Owned	Percentage Property Taxes Paid[a]	Tax Burden Ratio[b]
Top 1%	22.0	4.7	.214
Top 5%	31.3	7.1	.226
Top 25%	42.7	10.0	.234
Top 50%	47.4	11.5	.243
Bottom 50%	2.82	1.59	.564
Bottom 25%	.95	.81	.853
Bottom 5%	.13	.30	2.38
Bottom 1%	.02	.23	11.50

Source: Appalachian Land Ownership Study, 1980.
[a] Data from 1977 Census of Government for fiscal year 1976-77.
[b] The tax burden ratio is the percentage of property taxes paid divided by the percentage of surface land owned.

The Tax Burden

property-tax laws and practices is to shift the tax burden to the smaller owners, likely using land for homes and businesses, while leaving the large corporate or absentee owners of the surface, who likely are holding land for speculative purposes and can afford to pay, carrying little of the tax burden.

Even though the "poor pay more" while the property wealth of the region goes underassessed, the average county in Appalachia still does not generate adequate revenues for county services. In the eighty counties, only 22 percent of county revenues are raised from property taxes, while the average county in the nation as a whole gleans 31 percent of its budget from this source. For much of the rest of these funds, Appalachian counties must turn to federal and state sources. The average county studied received 49 percent of its revenue from nonlocal sources, while the average county nationally received 45 percent.[16]

Since the 1960s' War on Poverty programs, of course, the nation's taxpayers have poured federal and state funds into Appalachian counties on the assumption that the funds were needed to develop a depressed region. The irony of the federal and state subsidies is that they are going to the counties with the most valuable taxable resources. Overall, the counties with the highest coal reserves receive the most outside subsidy—58 percent of the revenues of the major coal counties comes from federal and state sources, compared to 49 percent for the sample as a whole and 45 percent nationally. In Martin County, Kentucky, 86 percent of the total county budget comes from intergovernmental sources, despite the fact that the county contains some of the most valuable coal properties in the nation, owned by large and profitable corporations. However, the land in Martin County is taxed at only $.39 per acre for surface and less than $.01 per acre for minerals underground. In the twelve eastern Kentucky coal counties, 70 percent of the county budgets comes from federal and state sources. Yet, if coal in the ground were taxed at rates comparable to other property using methods described earlier, the new revenue received would be $8 million, equal to 40 percent of the total revenues received by these counties from state and federal sources.

The net effect of these patterns contributes more to the tax inequities in Appalachia: funds provided in the name of aid to a poverty-stricken region serve, at least in part, to subsidize the property taxes of the region's large land and coal owners—who escape taxation. As a result of the underassessment patterns in the region, not only do the small local owners pay more, but other taxpayers, paying federal and state taxes, also bear an additional burden.

Despite the fact that small owners pay disproportionately to what they own, and despite the state and federal funds poured into Appalachian counties, a number of county governments face a revenue crisis. As a result of the lack of funds, needed services cannot be provided.

As is seen in Table 21, while the average county in the country pays $220 per capita for service delivery, in the eighty study counties the average per capita expenditure is $206. Because of differing reporting procedures, a more accurate picture is seen by looking at each state. In every state except

Table 21. Per Capita County Expenditures, by State and by Sample Appalachian Counties

	Average Per Capita Expenditure for State		Average Per Capita Expenditure for Appalachian Counties Sampled
Alabama	$ 96.84		$ 90.30
Kentucky	92.37		71.55
North Carolina	478.27		461.31
Tennessee	339.41		407.83
Virginia	539.79		402.26
West Virginia	58.41		49.07
National Average	$219.94	Sample Average	$206.20

Note: Data from 1977 Census of Governments. The large variations among states are somewhat due to differing reporting procedures. More accurate comparisons are therefore made within each state.

Tennessee, the per capita expenditure in the Appalachian counties studied is less than the per capita average for the state as a whole. In Kentucky and Virginia the contrast is particularly sharp: per capita expenditures in the southwest Virginia counties in the sample are 25 percent less than the state average, and in the twelve eastern Kentucky counties, they are 23 percent less.

One of the most important services affected by inadequate property taxation is public education. According to the 1977 Census of School Finances, 51 percent of school revenues in the nation came from county or parent government sources; 68 percent of the local funding for schools comes from the property tax, making "property tax revenue ... the most important single source of own source revenue" for school systems.[17] School systems across the nation face a financial crisis due in part to inadequate property taxation. The same crisis exists in Appalachia. However, the irony in many Appalachian counties is that school systems need not experience lack of funds, for as has been seen, the region contains valuable, taxable resources from which revenues could be drawn. Yet, case studies in this survey show time and again that school finances are often most lacking in counties with the most resources. Examples may be found from each state in the study:

Martin County, Kentucky, is now one of Kentucky's largest coal-producing counties, and yet 86 percent of its budget is derived from state and federal sources due to the inadequate property-tax base. The largest owner, Pocahontas-Kentucky, a subsidiary of Norfolk and Western Railroad, owns a third of the county's surface and 81,333 acres of mineral rights (equal to 55% of the county's surface). Yet its property taxes on its surface land are hardly enough to buy a bus for the county school system, and the $76 it pays on its mineral rights would not even buy the bus a new tire, to replace the wear it receives on the county's unpaved and rough coal-haul roads. As a result of lack of funds, education expenditures in Martin County per pupil are 24 percent below the state average and 43 percent below the national average. Other services suffer as well.

In Walker County, Alabama, the largest coal-producing county in that state, the twenty-eight largest landowners own over 65 percent of the mineral wealth in the county, yet contribute only $8,807 in property taxes on mineral rights. Of this, only $5,020 goes to education, not even enough to pay the salary of one schoolteacher terminated owing to lack of funds in the county. For the last sixteen years, the Walker County School System has had to borrow money in order for schools to open each fall. For the past nine years, owing to insufficient funds, the teachers in Walker County have been paid one to three weeks late each fall.

The pattern extends to counties outside the coalfields as well. In Swain County, North Carolina, where federal holdings account for over 80 percent of the land, and where, as a gateway to the Smoky Mountains National Park, millions of tourist dollars are also spent per year, the county cannot adequately support schools and other basic services. Despite a tax rate high for the area, the county is able to generate only around 30 percent of its revenue from local taxes. Intergovernmental revenues make up the rest of the budget. Because of the lack of funds, school facility construction has often been postponed: a sixty-three-year-old high school building was finally replaced in 1975.

Like Swain County, Morgan County, Tennessee, has a large amount of tax-exempt land—over 55,000 acres are owned by the state of Tennessee for a state prison, a park, and a wildlife area. The exempt state lands combine with poorly assessed coal, oil, and gas lands to leave little property-tax income for schools or other purposes. As a result, the tax rate of $7.55 per $100 value is, effectively, the second highest in the state. Still, funds are insufficient. Bus drivers have struck because of poor wages; school buildings are old and decrepit. In one school last winter students wore overcoats in class owing to lack of heat. Under threat by the state to close the schools, the already overtaxed citizens have passed a bond issue as a short-term solution.

The largest coal producing county in the state, Wise County, Virginia's, immense coal reserves are owned primarily by just ten companies, who control over one-half of the county's surface. Despite the county's mineral wealth, the school systems remain poor. In 1978–79, Wise County teachers were among the lowest paid teachers in the nation; the average annual teacher's salary of $11,506 was 24 percent below the national average. Conservative estimates (using the formulas presented earlier in this chapter) indicate that if the mineral reserves of the county were more adequately appraised, the new revenues would equal $1.25 million annually or 80 percent of the total taxes currently generated from real property in the county.

In Lincoln County, West Virginia, expenditures per pupil and average salaries are consistently below those of neighboring counties; the county's students yearly rank fifty-third or fifty-fourth out of fifty-five counties in test scores, and the school system has been under a court-ordered investigation due to its poor facilities and services. Yet, the county contains within it some of the most extensive oil and gas deposits in the region, with Columbia Gas alone owning over 270,000 acres of mineral rights in the county. A citizen's complaint against the undertaxation of these resources recently generated over half a million dollars in new revenue for the county, much of it going to the school system, but more funds are still needed.

How widespread is this pattern of impoverished school systems amid underassessed property wealth? What is the relationship between ownership patterns and school finance? Within states, for example in West Virginia, certain relationships have been found. As the West Virginia state report makes clear, low per-pupil expenditures and teachers' salaries as well as high dropout rates are most prevalent in counties with a high concentration of landownership. However, across states, the relationship is difficult to explore because differences among school finance systems hinder the gathering of uniform data.

What can be explored, however, is a broader relationship between landownership patterns and the median education level of a county's population. As we have seen earlier, the greater the concentration of land, the lower the taxes paid per acre. Where there is concentrated land ownership there might also be a shortage of property-tax revenues for schools. While a number of factors affect median education levels—family background, economic opportunities in a given county, outmigration—certainly a key element is the ability of a school system to provide quality education for its students.

With these assumptions, and aided by the relationships seen in the data, we might expect that where landownership is highly concentrated, then schools may be poor and educational attainment may be low. Where there is less concentration of land (and thus higher tax base) the quality of education might improve, and the educational level might also increase. When tested on the seventy-two rural counties in the sample, these expectations hold. In twenty-nine counties with a higher than average concentration of land, twenty-one or 72 percent had a lower than average level of education. By contrast, in the 43 counties with low level of concentration of ownership, only twenty-one or 49 percent had lower than average education levels. Put another way, of the 30 counties with high education levels, 22 (73 percent) were in counties with low levels of concentration of landownership.[18]

Admittedly, the above test is inadequate to test fully the impact of landownership patterns on school finances. Other factors may be at work besides the quality of the school system in defining the educational level of the population. Certainly an important element would be the nature of employment in the county. To determine the causal flow further, more precise analysis is needed. Nevertheless the point here remains: Concen-

The Tax Burden

trated land patterns, found to be associated with low property taxes, are also associated with a low education level of a county's population. One key may be the lack of necessary funds for quality school systems.

The paradox of ailing, underfinanced school systems amidst highly valuable property resources is only one of the many symptoms of inadequate property taxation in Appalachia. The larger the owner of the region's land, the less the proportionate taxes paid. Gross underassessment of mineral resources—the average tax per known ton of coal in the ground is 1/50th of a cent—adds to the lack of tax revenues. In many counties, massive federal or non-profit holdings also contribute to the fiscal crisis. As a result of the inequities of the property tax system, the larger owners—usually absentee corporations—go undertaxed, while federal and state subsidies are poured into these "needy" Appalachian counties to provide a minimal level of services. Even with the intergovernmental subsidies, impoverished schools and inadequate services continue amidst growing, relatively tax-free, exploitation of the region's resource wealth.

FOUR

Economic Development for Whom?

Appalachia has long been recognized as an area that is economically underdeveloped when compared to other regions of the country or to the nation as a whole. In spite of the development faith that was apparent throughout the region around the turn of the century, this century has not seen the development of a mature, stable economy within the region.[1] Even as it moves into the last two decades of the Twentieth century, the region still finds itself overly susceptible to the fluctuations of the national and global economy. The boom and bust cycles of the coal industry and their economic and demographic effects are well known. The economic effects of development in noncoal areas are less well documented, but there is increasing evidence that such areas are subjected to similar fluctuations, although perhaps less severe (e.g., the susceptibility of recreation-tourism areas to recession and energy shortages).

In the last two decades, many development agencies and policy analysts have maintained that Appalachian underdevelopment grows from lack of integration into the nation's economy. The strategy that flows from this school of thought focuses on the need to overcome the region's isolation through building roads and highways; on the need to provide seed capital for new industry; on the requirements of training the region's work force, and so on. However, the policies growing from these perspectives have not concerned themselves with matters of ownership of the region's land and resources.

This view has been increasingly challenged over the last decade by one suggesting that even with growing "integration" into the nation's economy, economic development may not occur; rather, economic underdevelopment is associated with the external control of land and natural resources, which limits diversified growth and removes the wealth from the region. From this perspective, widely articulated by Appalachian writers, Appalachia is sometimes likened to a "colony," a victim of the same forces of corporate exploitation that affect the Third World.[2] Through control of the region's land and natural resources, these forces prevent the formation of the indigenous financial control and other requisites for economic development. For development to occur, in this view, strategies must be developed that deal with the problems of ownership and control of land and mineral resources.

Studies of the early industrial development of Appalachia would seem to lend credence to the latter school. Whether we look at the general historical literature or specific case studies, the story is the same—massive investment by external interests for the purposes of exploiting the region's natural and human resources. The years of change at the turn of the century (1880-1930) began a process of concentrated control of land and natural resources, and of subordination to outside interests, that permanently altered the economic and cultural face of the region.[3] While the extent of this process varied from area to area, the attractions of vast virgin forests and massive coal reserves were powerful magnets for outside corporations, speculators, and entrepeneurs, who focused their initial investments on acquisition of land and resources. The next several generations would reap mixed benefits from the economic development thus set in motion.

Regardless of the part of Appalachia that we examine, whether coal or noncoal, the early economic development seems remarkably similar. In the Blue Ridge counties of North Carolina and Virginia, as well as in numerous counties in the Cumberland-Allegheny Plateau, the coming of railroads spurred the exploitation of timber resources until they were exhausted (see Swain, Watauga, Grayson, Wise, and Logan case studies). In some of these areas, the devastated land was latter "salvaged" by the National Forest Service. In the Cumberland-Allegheny Plateau counties the development of mineral resources (particularly coal) attracted immense amounts of outside capital (see Campbell, Mingo, Logan, Wise case studies). Population booms resulted that were to presage the waves of in- and out-migration associated with the fortunes of the coal industry. In many of these counties a pattern of absentee, concentrated corporate ownership developed that has become more or less permanent.

ECONOMIC DEVELOPMENT IN THE COAL COUNTIES

In the chapter profiling land and mineral ownership in Appalachia (Chapter II), we saw that corporate, absentee, and concentrated ownership are all evident in the major coal-bearing counties in the sample. What effects do such concentrated, absentee, corporate ownership patterns have on economic development in the coal counties? They involve the power to control economic change, the drain of wealth from the region, and the impacts of the single-industry economy which derive from these ownership patterns.

In general, even today the greater the concentration of land in an area, the greater the ability of a few owners to dominate the economic development. In Logan County, West Virginia, where nearly all the mineral wealth is concentrated in the hands of eleven corporations, local resident Roscoe Spence summed up the pattern: "By controlling land, they controlled the jobs; by controlling jobs, they controlled the payroll; by controlling the payroll, they once could control where people bought; by controlling where people bought, they could control profit on earnings. It was a stacked up thing. The effect of it is that people who control the land control everything." While the control may not be as absolute in some of these areas now as it was in the traditional company towns, the power of absentee corporate owners to affect the economic future of local communities is still massive. The entrance of multinational energy conglomerates into the coalfields of Appalachia has brought a new scale of capital investment, technology, and corporate power to the region. Control of resources development (and thus the local economy) is moved farther from the local or state level, at the same time that single corporate decisions can radically change the economic future of a county. Whether in the traditional company town or in the new era of oil-controlled coal, the basis of the power in the region remains the same—ownership of the land and its resources.

These ownership patterns, one should recognize, do not occur at random, but instead are concentrated where the resources are and where the greatest wealth of the region is to be found. In general, a greater degree of corporate control is associated with the greater reserves of coal, a greater production of coal, and with the most "value added" in mining.[4] In turn, the control of resources helps to create a dependency on mining jobs for employment, such that the greater the corporate control of land and minerals, the greater the percentage of the labor force employed in mining.[5] While the average coal county had 15 percent of its work force employed in mining, in a number of instances the figure was much higher. Examples are found in the case studies: in Mingo County, whose fate has always been linked to coal, 35 percent of the labor force is in mining (1976). In other counties with a high degree of corporate ownership, the figures are similar: Harlan, 38 percent (1974); Pike County, 34 percent (1970); and Wise, 25 percent (1977). Throughout the region, the control of land by a single industry brings with it control of jobs, helping to create dependency of workers and their communities both on the landholders who own the resources, and the employers who provide the jobs (often these may be one and the same.)

Accompanying concentrated corporate control in Central Appalachia is an absentee ownership that draws the wealth from the region. In 1884, a West Virginia State Tax Report warned that residents should become aware of the wealth of their minerals or "this vast wealth will have passed from our present population into the hands of non-residents, and West Virginia will be almost like Ireland and her history will be like that of Poland." Over time, that prediction has proven accurate. Like corporate ownership in the major coal counties, absentee ownership, particularly out-of-state ownership, is associated with the greatest coal production.[6] As a result, large amounts of capital leave Central Appalachia, according to a government report, and enter "the financial markets centered around New York" and other metropolitan centers.[7] Another indication of the drain of wealth is that a smaller portion of bank deposits in the coal counites studied are in time deposits (54 percent) than is the case in the noncoal counties (71 percent), suggesting that many deposits may merely be pass-throughs to other financial institutions outside the region.

Economic Development

Local planners, who are constantly faced with the problem of inadequate financial resources for development projects, recognize the outflow of wealth as a major problem. In the words of a planner in Harlan County: "Harlan is one of the wealthiest counties in the country, but not in terms of local capital or development. The money is not in Harlan banks, but in banks located in the eastern part of the United States." The loss of wealth to the absentee owners leads another planner in Pike County to observe, "there need to be controls on the amount of money absentee companies take out of the county."

Even within the region, however, there are numerous indicators that this coal-dependent economy is not one in which the maximum number of people benefit. While there is no doubt that a small number of indigenous residents have gotten very rich from the coal boom of the last decade, the wealth of these few regional entrepeneuers exists alongside considerable poverty and employment instability. In Pike County (usually touted in the media for its personal wealth and with one of the highest median incomes in the coal counties), 20 percent of the county's population had incomes below the poverty level in 1978. In Martin County, a current boom county, one-third of the population fell below the poverty level (1976); in Harlan, 25 percent were below poverty level (1978), despite the coal boom. And while average incomes have generally increased over the last decade owing to the coal boom, this tells only part of the story. These incomes (both per capita and median) are still usually less than the respective state averages. In 1977 Mingo was thirtieth of fifty-five in West Virginia in per capita income. For the coalfield counties surveyed in Virginia where corporate owners control almost one-third of the total land area, the average per capita income was only two-thirds of the state average; and the median family income was only 63 percent of the state average. Wise County, Virginia, demonstrates the apparent failure of the benefits of the coal boom to trickle down throughout the local populace. While per capita income increased between 1970 and 1977, the percentage of total personal income derived from transfer payments also increased substantially (from 15.6 percent of 19.4 percent).

An analysis of economic development patterns in coal counties of Appalachia must start, then, with several observations: the dominant single industry development is highly dependent upon the control of a few, primarily corporate hands, who control the land and resources; while large amounts of wealth are produced, much of it leaves the region. Even the wealth that stays in the region is unevenly distributed, leading to the persistence of poverty amidst riches.

In order to offset these patterns, economic development agencies such as the ARC have adopted a strategy of economic diversification. Counties like Russell County, Virginia, have taken a similar stand: "The area's leaders should do everything in their power to attract other industry, so that the area's economy is not so strongly tied to coal. The coal industry has a volatile history, and it is important that our dependency on coal is reduced." Individual residents affected by the lack of alternative opportunities often express the problem more poignantly. Says a Harlan County woman: "Mining will be the life of my three sons. If they don't mine, they can't make a living: either you mine coal or you push a buggy at Cas Walkers' [supermarket]."

Despite the fact that economic diversification is a widely expressed goal, nondiversification continues as the order of the day. The patterns can be seen by comparing the percentage of the work force in mining with the percentage in manufacturing, for select counties. On the average, in the major coal counties, 18.5 percent of the work force were engaged in manufacturing, compared with 28 percent for the overall sample. In some counties in the heart of Central Appalachia, the problem is more apparent. In Mingo County in 1976, 35 percent were employed in mining, while only 7.4 percent were in manufacturing. In Harlan County in 1974, 38 percent were in mining and only 5 percent in manufacturing. And in Martin County there are no manufacturing plants at all.

A number of reasons have been given by development agencies for the lack of economic diversification. These include isolation, topography, poorly trained work force, and lack of transportation and services infrastructure. The data suggest that the impact of landownership patterns must be included as one of the elements contributing to the lack of economic diversification.

The strongest indication of the effects of landownership patterns is seen in the proportion of the workforce engaged in manufacturing: the greater the corporate ownership, the lower the percentage of the workforce in manufacturing. Out-of-state ownership, too, evidently has a negative effect on the percentage of the labor force in manufacturing.[8] There is also a relationship between out-of-state ownership and the number of manufacturing establishments, such that the greater the out-of-state ownership, the lower the number of manufacturing establishments. And a similar negative association is found between out-of-state ownership and the value added in manufacturing.[9]

In the Virginia coal counties there is a noticeable absence of noncoal-related industries in counties most dominated by absentee corporate control of land and minerals. For example, Buchanan County, with a high level of absentee corporate ownership, had only three nonmining-related manufacturing establishments in 1976, whereas Tazewell County, with a relatively moderate level of such ownership, had fourteen nonmining related industries. While other factors may be operating in this differential, our regional correlations for coal counties indicate that absentee and corporate ownership are important contributing influences.

If landownership patterns do impede economic diversification, what are the mechanisms by which this happens? The two most prominent means seem to be: problems with the availability of land and the lack of an infrastructure adequate to attract and maintain diversified industry. In the words of the managing director of the Logan County Chamber of Commerce: "Logan County needs more industry, but the first thing they ask us when they want to come is if land is available. Then they ask about water and sewage. Of course, all of the answers are no." (Logan Case Study.)

In many instances, the interest of the large landowners seems to be simply in holding the mineral lands for speculation and future energy extraction, rather than in making them available for other forms of economic development. The effect is to keep land off the market and out of the local and regional economy, thus, among other things helping to insure their control of that economy. The extremely low taxes paid by the companies allow them to do this at little expense to themselves and with little contribution to local tax revenues. In Pike County, the impact is described by a former mayor of Elkhorn City: "This corporate ownership keeps the community from growing. As far as absentee owners, they don't spend no money in the county or in the state. I was raised next to Kentland's property, and they never did anything with it, just left it sitting. I know they've owned it for fifty years or more. They pay pasture taxes on coal-rich land. Where I grew up on Ferrell's Creek, Kentland owns the bottomland, big bottoms just sitting there."

Case studies report that land for industry and/or housing is often scarce in many counties, partly as a consequence of this continuing underdevelopment of vast areas of land. In Pike County, most of the coal-related corporations have not seen fit to sell their land for alternative industrial or commercial development. In Martin County, the landholdings of Pocahontas Kentucky, the dominant owner in the county, remains undeveloped except for coal mining. In Campbell and Claiborne counties, Tennessee, a local development group has been unable to obtain land for industry. In Harlan County, the expense of purchasing land with no improvements is prohibitive. In Mingo County, the only manufacturer of any size in the county is reportedly leaving because there is no land for expansion. Thus, in those case-study counties at least, the refusal of corporate landowners to sell their land for noncoal uses limits the areas in which commercial and housing growth can take place. However, while the availability of land is a necessary condition for industrial development, it is not a sufficient one. Several other factors also affect where and how development occurs.

Among the numerous factors considered by an industry in its decision on whether to locate in an area, the presence of an adequate services infrastructure is usually high on the list. Decades of absentee corporate ownership in the Central Appalachian coal counties have failed to produce adequate water, sewer, transportation, health, and educational facilities. This has come about for several reasons, only a few of which can be discussed here.

Certainly, one of the most obvious factors is that corporate-owned coal interests have not produced sufficient taxes to provide local revenues to develop such services. The minimal tax revenues received have hardly been adequate to meet the immediate needs of local communities, much less to provide the additional resources necessary for developing new services.

Past attitudes and behavior of large corporate owners have also played a critical role in the present condition of such services as water and sewage facilities. A former health officer in Logan County spoke of the tendency of some land companies to oppose sewage and water laws. In other instances large corporate holdings inhibit diversification by directly preventing the construction of such facilities. When such holdings are adjacent to urban communities, the result is often uneven development, since the construction of necessary facilities is restricted either to already built-up areas of the county or to more distant properties not owned by such companies.

In addition, the lack of locally available capital associated with absentee ownership minimizes the local funds available for housing loans, underwriting of industry and business, and construction of needed service facilities. In some counties landholding companies can effectively control the use of local capital through the placement of company or family representatives on bank governing boards or by obtaining controlling interest in a number of local financial institutions. The situation in Logan County is reported to be such that no capital projects can be undertaken without the sanction of one of the largest corporate owners in the county. The power and wealth of such companies often result in an arrogant disregard for the economic and social development needs of the localities in which they operate. A county planner in Pike County, Kentucky, refers to this as a lack of civic pride and speaks of the need to "force a little civic pride." A former health officer in Logan County, West Virginia, puts it more bluntly in his assessment that railroad companies "have historically operated on a public-be-damned basis." The net effect is summed up very aptly by a resident of Martin County: "These companies are taking their money out of the state and leaving nothing behind but wages: no roads, no recreation, nothing."

This history of one-industry dependence and its associated obstacles to industrial diversification have left most planners pessimistic about any chances of alternative economic development. Rather, the future is coal. There is almost an exuberant faith in the expansion of coal and its benefits. Local officials, coal industry representatives, and planners alike joined the synfuels bandwagon as they competed for liquefaction and gasification plants. This interest, however, seemed to be closely associated with the availability of federal dollars to underwrite such projects. Even regional planning units seem to be resigned to, if not enthusiastic about, the future of coal and the nonfuture of alternative industry. A planner with LENOWISCO, a planning district in the southwestern corner of Virginia (including Lee and Wise counties and the city of Norton), said simply that the agency did not see economic diversification as a realistic goal for Wise County.

While the faith in the promise of coal development is currently strong, the dependence on this single industry still heightens the degree to which the region is subject to a boom and bust economy. It is perhaps, too, the boom-and-bust cycle that helps to disguise the more permanent conditions of relative poverty of a large number of the population. When times are bad, they are bad for all; when they are good, the boom helps to cloud the fact thay they are still bad for some. In fact, booms, as well as busts, place strains

Economic Development

on local communities, which are aggravated by the patterns of concentrated landownership.

Booms may bring increases in jobs and wages, but they also have less positive effects, mainly those associated with rapid population growth, and increased demands for public facilities, housing, and services. For communities in which a diversified mature economy is already in place, there may be a capacity to absorb such rapid economic growth.[10] But for areas lacking such prior development, the strains are likely to be greater, and they may be intensified by ownership patterns. For instance, for a county already lacking available land for housing and public facilities, a rapid influx of population will place even more demands on existing stock, leading to overcrowding and rising prices. For counties historically plagued with patterns of underassessed corporate land, funds for providing new services are simply nonexistent. Schools become more overcrowded and roads overused.

Of course, the boom-town syndrome has long been a way of life in the region. During the first half of the century, many of the counties in the Central Appalachian coalfields experienced dramatic population growth, largely the result of rapid expansion in the region's coal industry. Now new proposals for the production of energy including coal-mine expansion and new synthetic-fuel plants indicate the possibility of a new boom period for many communities.

An example of how the already existing problems of "boom town" growth can be exacerbated by landownership patterns is described in the Wise County, Virginia, case study:

> Once a rural agricultural area, Wise County was rapidly transformed by coal industrialization at the turn of the century. The population of the county grew from 9,345 in 1890 to 19,653 in 1900 to 34,162 in 1910—a 266 percent increase in twenty years. With the growth, came a change in ownership and use patterns. Prices skyrocketed as speculators bought and sold land. By 1928, four large coal companies owned more than two thirds of the land area in the county. Land used for agriculture dropped rapidly: in 1860, four years after the county was organized, 196,606 acres of the county were considered farmland; by 1910 the farmland acreage had dropped to 122,848, by 1920 to 72,877, and by 1969 to 20,707 acres. In the 1930's and then in the 1950's Wise County was hit by a coal depression. With their land and agricultural base gone, without a diverse economy, people left the region. Population declined to 39,039 by 1971, the lowest level since before 1920.
>
> However, with the increased energy demand of the early 1970's Wise County was again faced with a coal boom. Population increased by 7,000 people between 1971-75. With land still tightly controlled and unused by coal owners, there was little room for economic or residential expansion. By 1975, 74 percent of the population lived in areas classified as "urban and built-up"—an

area constituting only two percent of the county's land area. While for the whole county, population density was only 111 persons per square mile, for this two percent of the land it was 4,035 persons per square mile, more crowded than the cities of Richmond or Roanoke. With the population increase, housing and other prices soared, the county experienced climbing crime rates, cultural disruption, and strained services. Now, the county faces the possibility of further population boom. There is a possibility of a synthetic fuels plant. However, according to a Department of Energy study, the construction phase of the plant could more than double the existing population, and the permanent population could increase by 4,600.[11]"

Not only does the concentration of ownership hinder adequate planning for economic growth, but the aura of corporate secrecy that often characterizes plans for economic expansion may also make matters worse for local officials. Given the scale of capital controlled by the contemporary, corporate owners of Appalachia, decisions about a single new mine or plant by a corporation can have major consequences for a local community. Yet, rarely are local officials or citizens given information for full planning to meet these contingencies. An example is found in Scott County, Virginia, in which the small community of Dungannon has been beset by rumors of a major new mine being opened by Consolidation Coal Company, a subsidiary of Continental Oil. Local citizens and officials have tried for some time to clarify these plans so that they can plan accordingly. They have met with little success and instead are faced with major uncertainties about future developments in the county. A county commissioner noted, "All Consolidated told one member of the board of supervisors is that until they decide to make an announcement, they won't say anything." In the same area, another firm is laying plans for the development of a large synthetic-fuels plant. But company representatives have refused to answer questions about the facility in public meetings.

Thus, in a manner reminiscent of previous boom-bust cycles, the public is left in the dark as to plans that will possibly precipitate a new boom period. They are once again left to the mercy of a coal-dependent economy manipulated by corporate interests beyond their control or influence. Given this dependency and their inability to influence corporate decisions, they are left to wonder if the projected boom is but another prelude to a bust for which they will bear most of the consequences. These busts can be devastating to the local community and its residents.

Dependency upon a single industry heightens the impact a bust can have on a local community. When the coal or energy market is down, unemployment is rampant; there are no other job options. Lacking the tax base, which in many counties is increasingly built upon the rate of coal production through the severance tax, communities and services suffer. Facing no other alternatives, people leave the area in search of employment and better community conditions.

Economic Development

Case studies and state reports in this survey illustrate the out-migration patterns, which occurred most dramatically during the coal decline in the post-World War II period (1950-70). During this period, the Big Sandy counties of Kentucky lost nearly 100,000 people to out-migration; the four survey counties in the Kentucky River area lost about 88,000 people; Harlan, Bell, Knox, and Laurel counties of the Cumberland River basin lost 100,000 people between 1950 and 1972 (Kentucky State Report). In Logan County, 34.3 percent of its population left following the coal slump of the 1950s; between 1960 and 1970 the population declined another 24.9 percent (Logan Case Study). The population of Wise County, Virginia, including Norton, reached a high of 56,336 in 1950, declined by approximately 14 percent to 48,592 in 1960, and declined again by over 17 percent to 40,119 in 1970 (Wise Case Study). For all the coal counties surveyed, the average rate of out-migration from 1960 to 1970 was 19.5 percent.

There are complex reasons, of course, why busts in the coal economy occur when they do: mechanization, the advent of strip mining, a changing market for coal—all were factors contributing to this particular decline. While land and mineral ownership patterns contribute to decisions on where and when coal is to be mined, they are only one element governing the boom and bust cycles of the coalfields.

The important point, however, is that concentrated landownership patterns limit the economic options that do exist when busts occur. Where landownership patterns limit economic diversification, few other jobs are available. With concentrated landownership, access for much of the population to the land itself is limited, even for tilling the hillside—a traditional mode of survival in the region. When a "bust" occurs, the likelihood of out-migration as the only option increases.

If this understanding is accurate, then we might expect that during times of decline in the coal market, coal counties with a higher degree of control of land resources will experience higher rates of out-migration than will counties where the land patterns are more diversified. In fact, for the coal counties surveyed in this study, there is strong positive association between the degree of corporate ownership in a county and the level of out-migration from 1960 to 1970.[12] There is a similar association between the level of absentee ownership, especially out-of-state ownership, and the rate of out-migration during the same period.[13] An example may be found in Harlan County, Kentucky, where 64 percent of the land is owned by corporate and absentee interests, and 38 percent were employed in mining. Between 1960 and 1970 Harlan County lost 36 percent of its population. In West Virginia, only one of the sample counties with a high concentration of large corporation and government holdings experienced a growth in population between 1950 and 1976, while nine lost population. McDowell, Logan, and Mingo, which have the greatest amount of this type of ownership, were among the top five in population loss, losing 48 percent, 38 percent, and 26 percent of their population, respectively (West Virginia State Report).

In the 1970s, changes in the migration patterns in the coal counties of Central Appalachia were brought on by a new rise in the coal market. For example, Mingo saw an 8.3 percent population increase between 1970 and 1976, accompanied by a decrease in the unemployment rate. Logan gained 1,000 jobs between 1970 and 1976 and showed a slight increase in population. Wise County, which in 1971 had its lowest population since before 1920, saw an increase of 6,000 people from 1970 to 1975. The counties in the eastern Kentucky river basins also saw population increases. But this reversal of out-migration in the coalfields is deceptive. There is no indication that the dependence on the coal industry has been altered or that a healthy, diversified economy has developed. If historical experience is any indicator, the current expansion of the coal industry will, as in the past, be subject to ebbs and flows. Indicators of such instability were already evident in some areas as the decade ended, when there appeared the anomaly of increasing coal production accompanied by decreasing employment in mining. For example, in West Virginia coal production increased 33 percent in 1979 to the highest level since 1973, while at the same time as many as 10,000 coal miners were out of work. Without economic diversification, without removing the dependency upon a single industry, the economic susceptibility is likely to continue.

ECONOMIC DEVELOPMENT IN TOURISM COUNTIES

The history of tourism counties differs from that of many coal counties. Their initial development at the turn of the century was not predominantly recreational, whereas energy development was clearly the future for the coal counties. However, the turn-of-the-century experience of what were later to become recreational counties was similar to that of the coal counties in that economic development was based on extractive industry. Just as coal and timber resources had attracted outside capital in the coal counties, the vast virgin-forest resources of the Blue Ridge and Allegheny Highlands attracted outside investment. This investment, coupled with the building of railroads into these hinterlands, was to spur enormous growth in the lumbering industry over the next few decades. Single-industry—often single-company—towns sprang up where nothing but wilderness had existed before.

This period of change, from the 1890s to 1920s, was a boom era for many of these counties, In Swain County, lumbering became a major industry in the early 1900s and continued so until the mid-1920s and the creation of the Great Smoky Mountains National Park. The population of the county grew from 10,412 in 1918 to 13,224 in 1920, the kind of surge representative of many such counties. Watauga County experienced a boom that lasted into the 1930s, by which time the timber resources of the county were largely exhausted. It was a time of relative prosperity, but the extractive basis of that prosperity and the timbering practices of the companies were ultimately to ensure its end. The timber that fueled the building needs of a developing nation was to provide few long-range economic benefits for its host counties.

Instead, the legacy was the virtual exhaustion of the area's forests, environmental devastation, and ghost towns, some of which were later to be promoted as tourist attractions. By the late 1920s the boom had run its course in most of the counties and the effects of the bust were readily apparent. Many of the towns built on the foundations of the timber industry were either reduced to rural villages or had disappeared altogether (e.g., the twin towns of Whitmer and Horton in Randolph County, West Virginia). With the exhaustion of the timber resources these towns had little economic base, nor were the railroads of any further value. They were instead to become relics, some to be developed as tourist attractions to supplement the later tourist appeal of the area (e.g., Cass Scenic Railroad, Tweetsie Railroad).

The exhaustion of the timberlands in the region also encouraged the entrance of a new type of ownership in the region—that of the federal government—which was to stimulate recreational development as the basis of local economies. It was in part the legacy of devastation that led to the acquisition by the National Forest Service of large acreages of "forestland" for purposes of timber management and preservation. One of the major impacts of this and other types of federal ownership (e.g., national parks and recreation areas) over the last several decades has been to encourage tourism and recreation, perhaps at the expense of other economic development. While there were certainly signs of the coming tourist/recreation industry already present, extensive federal ownership provided an incentive without which the history of recreational development would likely have been more gradual and less dominant in local economies.

While coal counties are dominated by corporate landownership, the tourist counties reveal a pattern of government and individual ownership. Government ownership accounts for 29 percent of the land sampled in these counties, three times the level found in the high coal counties and almost double the level found in the agricultural counties. As one might expect, there is a strong correlation between the percentage of government ownership in a county and the level of recreation and tourism development.[14]

Despite the federal presence, individuals still own 48 percent of the land in the sample. However, many of these are absentee owners who likely are holding land for speculation or second homes. In some recreational counties, the level of nonlocal individual owners has increased dramatically in recent years, as tourism and recreation have become increasingly the basis of local economic development. This trend was documented in a study by the North Carolina Public Interest Research Group, which noted that from 1968 to 1973 the total number of acres held by local residents in their ten-county study area dropped by 10 percent, while nonlocally owned land jumped from 28 percent to 36 percent of all private land.[15]

The combination of land held by absentee individuals and the federal government in the tourism counties leads to a level of absentee ownership comparable to that of the coal counties. And the degree of control of land in the tourist counties by all of the absentee, government, corporate, and large individual owners in the sample is even greater than in the coal counties. In the tourist counties, those interests control some 60 percent of the total land surface.

At first glance, the post-World War II economic experiences of the recreational counties have not been characterized by the extremes that affected the coal counties. Even though some of the recreational counties experienced something of a bust surrounding the 1974 energy shortages and recession, most have not had the dramatic population fluctuations of the coal counties. For instance, western North Carolina counties continued to gain population during the 1950s and 1960s, contrary to the trend in the coal counties. This was probably due to several factors: the presence of small-scale agriculture, the absence of extractive industry dominance, and a somewhat improved and more diversified economic situation. Watauga County actually experienced a population increase of 33.5 percent from 1960 to 1970, precipitated by the growth of Appalachian State University and the recreation industry.

If population growth is used as an indicator of economic growth, it would appear that the situation in these counties improved dramatically during the 1960s-70s. Or if the rate of employment growth is seen as a sign of economic growth, counties like Watauga (with a rate of employment growth three times population growth during 1960-73) would seem to have developed very healthy, dynamic economies. However, in many of these counties a familiar pattern was emerging—that of one-industry dominance. Spurred on by federal forest ownership, the promotions of state, local, and regional agencies, and the proximity to vast urban populations, the recreation industry began to experience phenomenal growth and to dominate other sectors of the economy. In counties like Watauga and Avery, a rapid increase in such development over the last twenty years brought with it a surge in second-home and resort developments. (See discussion in Chapter 6.) A new economic dependency was in the making, which, like those in other areas of Appalachia, meets the needs of outsiders at the expense of local residents.

The subsequent economic development has been neither diversified, nor stable. Nor has it in most instances lived up to the rosy predictions of its supporters. In Grayson County (an agricultural county slated for recreational development), the predictions of a local leader that Grayson Highlands State Park would bring in 200,000-500,000 people per year has proven illusory. For the 1979 season there were 18,000 visitors, approximately half the total for 1978. Yet, in spite of such experiences and numerous studies that have questioned the advisability of recreational development, regional planners seem to have maintained their enthusiasm for it.[16]

The impact of recreational development on the economic situation of area residents can be examined in several ways: the types of employment it produces, the development it encourages in other economic sectors, and the development it impedes in other sectors. There has been much disagreement about the overall economic benefits of recreational development for local residents but findings tend to corroborate that "for the majority of the people the economic impacts are more negative than positive."[17] This would confirm ARC's preliminary investigations into the impact of tourism and recreation in Appalachia, which warned that the resort industry is "low pay and seasonal in nature."[18]

For instance, in Swain County in 1976, 23 percent of the total labor force was engaged in travel and tourist-related employment. This, coupled with manufacturing employment in low-wage textile and furniture industries, produced a per capita income in 1977 of $4,368. Only sixteen other counties in North Carolina recorded lower figures for the same period. While the employment rate has grown considerably in Watauga since the mid-1960s, it does not seem to be reflected in increased wages and income, since much of the growth has been in low-wage and seasonal employment. In 1973, for example, the county's per capita income was only 73 percent of the state average. In 1976 the average weekly wage was only 76 percent of the state average, further indication of a low-wage economy. Another county experiencing the seasonal and low-wage employment of resort and recreational development—Cumberland County, Tennessee—had a per capita income 67 percent of the state average in 1977.

Given such considerations, one must question the promises of recreational development as a strategy for economic resuscitation, a rationale given for the Mount Rogers National Recreation Area in southwestern Virginia. One of the counties projected to benefit is Grayson, until recently a predominantly agricultural county. In 1950, when employment was primarily in the agricultural sector, the average weekly wage was 83 percent of the state average. In 1977, after a significant shift away from agriculture, it was only 58 percent of the state wage. The proposed national recreation area is touted as a means of improving this once it is fully developed. Yet, in the environmental-impact statement for this development, the projected annual payroll is $12,637,736 for 3,272 people or some $3,862 per employee, hardly an annual salary likely to increase either weekly wages or per capita income.

Unemployment and cyclical employment are also the fruits of a tourist-based economy. For the high tourist counties in our sample, the average county experienced an unemployment rate of 7.74 percent in 1977, slightly higher than the figure for the average coal counties (7.34 percent). Within the tourist counties, the ownership of land by government, absentee individuals, and corporations (most of which are involved in resort development or forestry) is associated with unemployment, such that the greater the percentage of a county owned by these interests, the higher the unemployment rate. High concentrations of ownership in these counties, usually caused by large blocks of federally owned land, shows an even stronger association, such that the greater the concentration of landownership, the higher the unemployment in the recreation counties.[19]

Looking at particular recreation counties, Swain had an unemployment rate of 9.9 percent in 1977; Cumberland a rate of 10 percent in 1979. Watauga County usually has an unemployment rate higher than that of the state except in the summer months, when it is lower due to increased recreational employment. The specter of underemployment, which is not indicated by these figures, is perhaps even more important. The low wages, cyclical employment, lack of high-skill jobs, and high rates of participation in social assistance programs would lead us to believe that the rate of underemployment is quite high.

These conditions of unemployment and underemployment exist at the same time that the tourist-based industry brings with it a higher cost of living for area residents. Once again, Watauga serves as an excellent example: it has ranked seventh or higher out of North Carolina's one hundred counties over the past several years in cost of living, while ranking as low as seventy-ninth in per capita income. The implications of such a situation for local residents should be obvious, particularly when accompanied by increased housing and land prices brought about by real-estate speculation.

The economic underdevelopment found in recreational areas also results from the character of secondary development that the tourism industry spawns. Not only are the jobs in the recreation industry menial and low-paying, but so also are those in the retail and services sectors that support it. Significant growth has occurred over the past several years in the retail and service components of counties such as Watauga. In Watauga, employment in the hotel and lodging segment of the economy is 6.4 times greater than that of the state as a whole. However, much of the employment in these sectors is both low-skill and low-wage, many jobs paying minimum wage or less. Additionally, the wages paid in the trade and service sectors in Watauga were well behind those of the state (74 percent of the state average in trade and 84 percent in services). The picture rapidly becomes one of a low-wage economy in a high-cost environment.

The manufacturing sector in the recreational counties is critical to economic diversification. It is difficult to associate the presence or absence of manufacturing facilities with the availability of land in these counties—on the whole, land is not as tightly controlled as in the coal areas. There are exceptions, however, such as Swain County, which has a tourism and low-wage service-industry base in which most of the population is employed in nonmanufacturing jobs. The extensive public ownership in the county (80 percent) has apparently affected the availability of suitable land for industrial development, since most of the remaining level land in the county is within public boundaries, and thus unavailable for industry. An interviewee stated that graded land elsewhere in the county costs so much as to be prohibitive ($75,000 per acre). Availability of reasonably priced land for housing could also pose a problem for attracting industrial development in recreational counties (see discussion in Chapter 6).

In another respect, the low wage levels and cyclical employment in the recreation industry make it possible for traditionally low-wage manufacturing establishments in the area to remain so. In fact, the presence of manufacturers in these counties (associated positively with corporate ownership) does not seem to have a positive effect on income levels. There is, for instance, a negative association between corporate ownership of land and per capita income, and a positive one between such ownership and the

Economic Development

percentage of families below the poverty line.[20] Thus, while there may be the impression of economic diversity in some recreational counties, it is a diversity based on low wages and unstable employment.

One other element affecting economic diversification in the coal counties also appears important in recreational ones—the availability of local capital. As mentioned in discussions in the coal section, local capital is necessary for infrastructure development, land purchases, building activity, loan making, and so on. The problem in the coal counties is that great amounts of locally derived wealth are shipped elsewhere, owing to absentee control of resources. In recreation counties, the story is different. Absentee ownership seems to be associated with a lack of local capital altogether; it seems to create little wealth to be expropriated.[21] Rather, the individual absentee ownership that predominates in recreational counties is for purposes of either personal aesthetic enjoyment or speculation, neither of which create much local capital. Likewise, government ownership is unlikely to produce the kind of local capital conducive to nonrecreational industrial and commerical development.

In sum, we find land ownership patterns contributing to one-industry economies in both coal counties and recreational ones. The appearance of economic diversification in the latter is deceptive, because the low-wage, seasonal employment created to service the recreational/tourism industry is overly dependent on the fluctuations of that industry. Whereas absentee corporate ownership is critical in the maintenance of a one-industry economy and economic underdevelopment in the coal counties, it is government ownership and the individual absentee ownership it encourages that seem to be most influential in the recreation counties. The results are similar in that industrial diversification is made more difficult by the lack of available land, inadequate local capital, and local tax revenue insufficient to provide an adequate services infrastructure.

FIVE

Appalachia's Disappearing Farmland

Appalachia historically has been thought of as the land of the small farmer. Studies by the United States Department of Agriculture (USDA) in 1930 concluded that the southern regions of Appalachia had the heaviest concentration of self-sufficient farms in the country.[1] Even today, many Appalachians share a closeness to the land, a familiarity with and attachment to it. Yet throughout this century, Appalachians have witnessed a constant assault on their land, resulting in the displacement of hundreds of thousands of small farmers and the disintegration of the culture and communities of farming.

Well over a million acres of farmland went out of agricultural production in the eighty counties of this study between 1969 and 1974, the latest years for which figures are available. Over 17,000 farmers left farming in this period—about 26 percent of the farming population of these counties. If these rates continued throughout the 1970s, the new Agricultural Census will show that in a single decade over half of Appalachia's farmers will have ceased farming and over a third of the region's farmland will have gone out of production.

The decline of the small farmer is, of course, a national phenomenon. In the late 1930s there were over 6.8 million farmers in the United States, all but a small percent classified as family farms.[2] Today the number is 2.3 million and still dropping. It is estimated that ten farmers a day leave the land. Total land in farms declined 2.35 million acres during 1979. The reasons for the loss of over 4 million family farms in this country since 1930 are complex and may vary in importance from region to region. The more significant factors appear to be the economic instability of small farms, the corporate intrusion into agriculture that has been aided and abetted by federal policies, and loss of land for agricultural use.[3]

At the heart of the small-farm crisis lies the economic disadvantage of the small farmer. New-style agriculture, with its intensive use of chemicals and machinery, requires a degree of capitalization that is often beyond the reach of small farmers. The small farmer feels the pinch from corporation "input" suppliers (machinery, feed, fertilizers, and seeds) and from the "output" corporations (the middlemen) that process, market, and retail the farmers' product. In 1974, the farmer received only forty-one cents out of each dollar the consumer spent on food. Only 6 percent of the rise in food prices between 1954 and 1974 went to the farmer. Moreover, a high degree of actual farm production is coming into the hands of corporate interests. This has occurred primarily through contract farming, which soon may account for over half of America's food supply.

A number of governmental policies have worked to the advantage of corporate and large growers and have given impetus to the disappearance of the small farmer. The most important of these special advantages are (1) agricultural support programs that subsidize the corporate interest in agricultural production; (2) tax laws (e.g., inheritance taxes) that place family farmers at a competitive disadvantage because of the variety of income-tax loopholes available to large, corporate farm units and nonfarm investors in farmland; (3) agricultural labor policies that work to the disadvantage of the small farmer; and (4) the research orientation of the USDA and the land-grant colleges, an orientation that has helped to develop the highly mechanized, capital-intensive pattern of production that has spurred the decline of the small farm.

Such factors, however, are not the only significant elements behind the farm crisis. As discussed in Chapter 1, the loss of agricultural lands to nonfarm owners has also been an issue of national importance. In Appalachia, too, patterns of landownership and land use contribute to the lack of land for agriculture. In general, corporate, absentee, and concentrated ownership patterns are each associated with a low use of the land for farming. Where such patterns are prevalent, or are newly emerging, agriculture competes with other land uses, especially energy and tourism development, bringing further pressure on the farmer. Combined with the other economic pressures on the small farm, patterns of landownership and use may encourage existing farmers to give up farmland, as well as discourage or prevent new farmers from obtaining it for agricultural production.

Originally, settlers came to Appalachia to hunt, to fish, and to farm a little.[4] The soil was rich and settlers turned more and more to raising corn and livestock. They used a primitive but productive style of agriculture—based on the Native American example—the slash-and-burn method. By the mid-nineteenth century, Appalachians had come to support themselves by means of subsistence agriculture, supplemented by an outside income raised first through hunting and lumbering and later by the sale of whiskey.

Corporate acquisitions by lumber and coal interests and the subsequent exploitation of coal and timber at the turn of the century limited the amount of land available to the Appalachian farmer. As a result, farmers were often left to farm land that they had never intended to use as their sole means of support. With this intrusion began the decline of mountain agriculture. In Wise County, Virginia, the site of the opening of the southwestern Virginia coalfields, there were in 1880, 1,145 farms covering 273,654 acres. By 1920 the number of farms had dropped only slightly to 1,067, but the land in farms had been dramatically reduced to a mere 72,877 acres, less than one-third the original area. The development of the national forest, prompted by the tremendous devastation of the region's woodlands, later played a similar role in shaping the course of subsistence agriculture in the moutains. For example, in 1911 the initial purchase unit (Whitetop) of the Jefferson National Forest in southwest Virginia was 11,358 acres; by 1978 its holdings totaled 683,675 acres.

The loss of land for farming in Appalachia, which began over a hundred years ago, continues through some of the same agents today. The timber industry and the coal industry have been expanding and consolidating their control over land in Appalachia. The expansion of federal holdings, begun in the second decade of this century, and the recreational development usually associated with it, add further to the pressures on agricultural land.

Beginning in the 1870s, the national need for lumber brought agents of timber corporations into Appalachia. They conducted title searches, which often led to Appalachian farmers' being stripped of much of the land that had supported them. As Harry Caudill points out, the Appalachian subsistence farmer usually titled only the small portion of the land that he actually cultivated, and, as a result, lost to the timber companies the untitled land where he had hunted and fished.[5] Farming was made even more difficult by severe siltation and flooding problems from the timber industry's logging practices and its removal of the region's virgin timber.

The development of the coal industry prior to the turn of the century led to the next major disruption of the land used by the subsistence farmer. The agents of the coal industry used various maneuvers to cajole Appalachians to sell their mineral rights. The result, in Caudill's words, was that the Appalachian farmer came to be "little more than a trespasser upon the soil beneath his feet."[6] Many subsistence farmers deserted their ancestral farms to take jobs in the coal camps, but a majority stayed behind to follow the same pattern of agricultural life. Dean Pierce describes what happened next:

> Those who remained on the land attempted to provide more food or whiskey to meet their own increased needs and the demands of the coal camps. The additional foodstuffs raised to sell to these camps led to the eventual and everlasting destruction of the soil. It was these increasing outside pressures that came to overstress the agricultural system and finally to destroy the fertility of all the soil. Moreover, the coal camps, through an unjust control of tax assessment, passed the tax burden back to the landowners, falling heavily upon the subsistence farmer, who could ill afford to pay for the area's desperately needed services.[7]

By the 1930s the Appalachian farmer had become so dependent on the coal industry's cash economy that he was totally unprepared when the Depression forced him once again onto subsistence agriculture to support himself. In Alabama, small landholders across the state were often unable to pay even their low property taxes. As a result, from 1928 to 1933, over 2.6

Disappearing Farmland

million acres of land in the state were sold for taxes out of over 41 million acres of land that were tax delinquent. Eighty-four percent of the land that was sold for taxes was farmland. Much of that farmland was purchased by large, land-extensive corporations, primarily timber companies.[8]

Those who had left their farms to become miners fared little better. In 1932, a survey of 956 unemployed miners in Kentucky and West Virginia found that only 11 percent wanted to return to mining, while 48 percent wanted to return to farming. However, by now the return to farming was blocked, for the miners no longer owned the land. Malcolm Ross, a *New York Times* writer, wrote in 1933 about miners who "would desire to return to cultivation of the land; the trouble is they no longer have any claim to it. The coal companies own the land."[9]

Historically, the Appalachian small farmers have clung to and fought for their land against very difficult circumstances. They continue to lose the battle.

Today, Appalachian farmers have much in common with small farmers elsewhere. They suffer from the same governmental neglect, financial instability, and corporate dominance that plague small farmers throughout the country. Yet there are some obvious differences. Appalachian farmers tend to be older, less educated, and poorer. The average farm in Appalachia is smaller, and the uneven topography results in the division of available cropland into such small and scattered fields that efficient use of machinery is at times impossible.[10] The pressures on farmland from energy development, tourism, and federal acquisitions pose special problems for Appalachian farmers.

One clue to the reason for farmland loss in Appalachia is found in Table 22. The greatest percentage of loss of farms in the survey counties was in Kentucky and North Carolina. The greatest percentage of loss of farm acreage was in Kentucky, West Virginia, and North Carolina. These are also the states in which coal or recreation developments have been greatest. In fact, all but one of the counties in the sample that lost 30 percent or more of their farms between 1969 and 1974 were significantly affected by tourist and second-home development or by coal production. (See Tables 25 and 26.)

Table 22. Loss of Farms and Farmland in Eighty Appalachian Counties, 1969–1974

State	Farms Out of Production	Acres Out of Production
Alabama	5,696 (25.3%)	442,578 (17.4%)
Kentucky	1,406 (31.4%)	118,531 (27.3%)
North Carolina	3,680 (31.3%)	205,056 (22.2%)
Tennessee	1,686 (20.4%)	163,388 (15.2%)
Virginia	3,183 (22.8%)	182,255 (11.5%)
West Virginia	1,366 (26.4%)	219,380 (23.9%)

Coal development in agricultural areas, especially strip mining, frequently destroys the land for subsequent farming, through acid mine drainage and flooding. The absentee corporate ownership associated with coal development limits future agricultural use of the land, since mineral lands are usually held for long-term speculative development. In the traditional coal counties, the barriers to housing and commercial development posed by corporate and absentee landholding in many areas have led to urban sprawl along the narrow river bottomland that is the major farmland in such areas. When large blocks of land are taken out of the housing market, farmland is often converted to residential development. Even a predominantly rural state like Kentucky lost 123,181 acres of prime farmland to urban sprawl from 1969 to 1979. While much of this loss was in areas surrounding the urban areas of central and northern Kentucky, Pike County, in the heart of the eastern Kentucky coalfields, was among the top counties in the state in terms of such loss.

Federal and state ownership, with its associated recreational development, has placed undue pressures on farmland in western North Carolina, southwestern Virginia, and elsewhere. When these acquisitions are accompanied by corporate purchases of vast acreages for purposes of building pumped-storage facilities and other dams to produce electricity, the loss of farmland can be significant. In areas where these ownership patterns are found in combination, land speculation can lead to a rapid escalation in prices for farmland, making either the retention or expansion of farmland more difficult. G. Halsey of the Grayson County (Virginia) Agricultural Stabilization and Soil Conservation Service gives an excellent example of the resulting price spirals. "Grayson Highlands State Park, Mount Rogers National Recreation Area, and APCO [Appalachian Power Company] all three buying land in the county at the same time caused the price of land to get higher. Countywide, land is now selling for $600-700 per acre, which is probably triple in price since the 1960s."

It is not surprising that the Appalachian farmer is older than average, when spiraling land prices have made it next to impossible for new or young farmers to begin farming. If an individual has not inherited a piece of land, the initial investment for land and operating equipment can be close to $400,000.

The striking loss of over a million acres of farmland with over 17,000 farmers in the sample counties of Appalachia between 1969 and 1974 is in part connected with the reasons for the national decline in agriculture during this period. Evidence also suggests that the landownership and land-use pressures discussed above contribute to the decline of farmland in the region. In general, there is a significant correlation between absentee and corporate control of land and the use of land for farming.[11] Two developments in particular, energy and recreation development, have had major impacts on the loss of farmland.

In Appalachia, corporate control of agricultural land does not seem to lead to agribusiness (corporate agricultural production) as it does elsewhere in the country. Here, corporate ownership takes land out of agriculture altogether. In our survey counties, the greater the corporate control of land, the lower the percentage of land devoted to agriculture.[12] Of the thirty-one rural counties with a higher-than-average amount of land in agriculture, 87 percent have a below-average level of corporate ownership. Of the twenty-six counties with a high level of corporate ownership, on the other hand, only four also have a high degree of the county devoted to agriculture.

Absentee ownership of land is also associated with low use of land for farming, as is concentration of ownership (greater acres in fewer hands). These associations suggest that where landownership becomes concentrated in a few corporate and absentee hands, it may be valued for reasons other than its farm potential (e.g., energy development, mineral and timber resources, recreation).[13] Farming of that land, even while it is not being otherwise used, will be discouraged. Indeed, we found that the less the local individual ownership, the less the use of land for farming and the lower the value of agricultural sales in a county.[14] This is illustrated in Tables 23 and 24. Of 28 counties with a relatively high level of land *not* owned by local individuals, 74 percent had a lower level of agricultural use in the county and 76 percent had a low level of agricultural sales. On the other hand, of thirty-four counties with a lower proportion of land not owned by local individuals, 62 percent had a high level of farmland and 62 percent had a high level of agricultural sales.

When large blocks of land are essentially taken out of local use because of their ownership patterns, one may expect the consequent pressure on remaining agricultural land to be great. Housing and economic development uses compete with small farmers for the use of the remaining blocks of available land, and both the consequent price spiral and related property tax pressures exacerbate the problems for family farms (see Chapter 3 on tax problems).

Table 23. Correlation of Ownership with Agricultural Use of Land

Percentage of County Not Owned by Local Individuals	Percentage of County in Agriculture		
	Low (less than 25%)	High (25% or more)	Total
Low (less than 40%)	13[a] (38%)[b] (32%)[c]	21 (62%) (68%)	47%
High (40% or more)	28 (74%) (68%)	10 (26%) (32%)	53%
Total	57%	43%	100%

Pearson's R = −.462 at the .0001 level of probability.
[a] Number of counties.
[b] Percentage not owned by local individuals (row).
[c] Percentage in agriculture (column).

Table 24. Correlation of Ownership with Agricultural Sales

Percentage of County Not Owned by Local Individuals	Level of Agricultural Sales		Total
	Low (less than $5 million)	High ($5 million or more)	
Low (less than 40%)	13[a] (38%)[b] (31%)[c]	21 (62%) (70%)	47%
High (40% or more)	29 (76%) (69%)	9 (24%) (30%)	53%
Total	58%	42%	100%

Pearson's R = -.437 at the .0002 level of probability

[a] Number of counties.
[b] Percentage not owned locally (row).
[c] Percentage agricultural sales (column).

Where farmers are unable to expand or improve their farms by acquiring more land because of high prices or unavailability of land, and where taxes are high, farmers may have to turn to other occupations to supplement their farm income. One may expect such a pattern to emerge more clearly in areas where farming is still practiced than in areas where it has been virtually eliminated. In the average county of our sample, 55 percent of the farmers gained more income away from the farm than on it. In the high-agriculture counties, fewer farmers held other jobs. But within those high-agriculture counties, there is a correlation between degree of absentee corporation and government ownership and the proportion of farmers with other jobs. The greater the absentee corporations and government ownership, and the greater the concentration of land in a few hands, the greater the percentage of farmers with other major occupations.[15] For instance, of thirteen agriculture counties in the sample with a higher than average level of concentration, twelve also had an above average percentage of farmers with other jobs.

In the general sample, some significant relationships have been found between landownership patterns and the structure of agriculture. Such relationships emerge even more strongly in two particular types of counties: those in which recreation and tourism are placing increasing pressure on the land, and those in which energy development is taking place.

AGRICULTURE AND LANDOWNERSHIP: TOURISM COUNTIES

Traditionally, agriculture has played a significant role in the economies of most of the counties classed as recreational. It continues to contribute

Disappearing Farmland

substantially to these counties' cash receipts—in 1976, in the twelve counties of the study in western North Carolina, cash receipts generated by agriculture amounted to $105,852,000. But the dynamics of tourism development threaten the continuation of agriculture as an integral part of many local economies. In particular, the pressure on farmland created by second-home development and resorts may destroy what was once the most stable element in a diversified local economy.

Case studies illustrate the trend. In Swain County (North Carolina), 26.2 percent of the county's land was in farms in 1939 (even after the federal government had made its major acquisitions for the Great Smoky Mountains National Park and the Cherokee reservation). These and subsequent federal acquisitions in the county have created a situation in which over 80 percent of the land is owned by the federal government. As in the case of many other western North Carolina counties, this ownership has spurred the purchase of second homes and recreational development. The combined effects have led to a dramatic decline in farmland in Swain County such that by 1974 only 2.8 percent of the land was in farms. According to one local resident, "There really hasn't been a young person getting into farming lately because of high land prices and outside pressure of people coming in from outside the county who are willing to pay a high price for [the land]. This has taken good land out of agricultural use and out of production."

In the five-year period from 1969 to 1974, the most recent for which data are available, high losses of farmland were recorded for many recreation counties (See Table 25.)

Table 25. Tourism/Second-Home Counties with High Loss of Farmland, 1969–1974

	Percentage Loss of Farms	Percentage Loss of Farm Acres	Loss of Acres	Percentage Tourism Services[a]
Swain, N.C.	46.9	27.4	3,700	85.5
Jackson, N.C.	40.3	41.6	15,175	43.3
Mitchell, N.C.	36.9	32.3	17,308	—[b]
Clay, N.C.	36.5	35.3	10,727	—
Randolph, W.Va.	36.5	25.6	46,442	24.5
Ashe, N.C.	35.9	19.2	33,010	12.5[c]
Avery, N.C.	35.8	24.9	10,352	54.6
Cumberland, Tenn.	34.3	14.3	15,820	37.4
Watauga, N.C.	30.3	16.2	12,338	64.0
Madison, N.C.	30.1	27.9	46,117	—

[a] Percentage of service receipts in the county based on hotels, motels, trailer parks, campgrounds, amusement, and recreation (based on 1972 Census of Service Industries).
[b] While data is not available for these counties, it is known from other sources that these counties are strong in tourism and second-home developments.
[c] While the tourism industry is not as high in Ashe County, other data indicate the number of second-home purchases to be high.

For most of these recreation counties, farmland loss was considerably higher than for the average county in the sample. Indeed, in three North Carolina counties (Jackson, Mitchell, and Clay) around twice the average farmland loss occurred. Recreation counties in West Virginia and Virginia were not far behind. In the eight North Carolina counties alone, almost 150,000 acres of farmland were lost in just five years, and over 2,700 farms —more than a third of the farms in these counties.

Landownership patterns have played a major role in this declining agricultural economy in the tourist counties. Second-home and resort development create land speculation and a price escalation that puts land prices far above what the local market can bear. Land values in relatively undeveloped agricultural townships of Watauga County (North Carolina), for example, increased an average of 225 percent in the twelve-year period from 1963 to 1975. Ross Payne, a local real-estate agent in Cumberland County (Tennessee) said that the general price of land has gone from $100 per acre, the price of land when he first came to the county fifteen years ago, to around $1,000 per acre now.

High land prices affect agriculture in several ways: They may tempt people to sell, and thereby put land out of agricultural use. They act as a barrier to expansion of farms or to new farmers entering the occupation (unless they have been fortunate enough to inherit a plot of land). Property taxes soar to meet new services demanded by the tourist economy. The increasing property-tax burden, especially hard in counties where much of the land is taken out of the local tax base by public and nonprofit ownership, increases the economic problems of "making it" in farming that already exist at a national level.

In these recreational counties, it is absentee and public ownership of land that has the major impact on farming. In many of the recreational counties, federal government ownership of land increases the pressure on and competition for already scarce land. Of the nineteen tourist counties, twelve have a high degree of public ownership. The average tourist county has 14.2 percent of its land in public ownership, almost double the average for nontourist areas of the sample. Of the remaining land in the county, out-of-state individuals own an average of 17.5 percent of the surface, compared with 12.2 percent in nontourist areas. Altogether, fourteen of the nineteen tourist counties (74 percent) have a high degree of absentee ownership.

Within the tourist counties there are also associations between these landownership patterns and other indicators of a farm crisis—such as farmers turning to other jobs, and increasing age of working farmers. One might expect that where farmland is being lost and farms are disappearing, farmers will not be able to maintain the economic viability of their operations. Not only will individuals cease to be farmers altogether, but also individuals who continue to farm will have to turn to other, nonfarm jobs, in order to supplement their farm incomes. Within the recreational counties, more farmers are turning to other occupations than in the agricultural counties of the sample (although to a lesser extent than in coal counties). Within the tourist counties in particular, the percentage of farmers taking other jobs is associated both with the degree of public ownership of land, and the combination of corporate and public landownership levels. Of the eleven tourist counties with a high level of combined absentee, corporate, and government ownership of their land (that is, with a lower-than-average level of local ownership), nine (92 percent) also have a higher-than-average number of farmers in other jobs.[16]

The pressures on the farm economy created by landownership patterns —land scarcity and high land prices in particular—limit the economic viability of farming in recreational counties. This not only increases the likelihood that people currently farming will seek other employment, but it also diminishes the likelihood that new people will enter farming. While the reasons that fewer people are choosing agriculture as a career are quite complex, but the study indicates that the scarcity of reasonably priced land may be a factor in the recreational counties. In those counties there is a strong correlation between the percentage of increase in average age of farmers (1969 to 1974) and the degree of public ownership of land in the county, as well as the degree of absentee concentrated ownership (such that fewer people hold greater amounts).[17]

What appears to be occurring today in the recreation areas of Appalachia is a process similar to that which occurred decades ago in the Central Appalachian coalfields. There, with the development of an energy industry, the people were displaced from their land and turned into the miners needed for industrialization. Today in recreation and agricultural areas, people are also being displaced, often to provide cheap labor for industries in the process of again industrializing the region, or for support services necessary for recreational development.

AGRICULTURE AND LANDOWNERSHIP: ENERGY COUNTIES

In the sample counties generally, the greater the level of coal production, the less the number of farms in a county; the less the farm acreage in a county, and the smaller the proportion of the county in agricultural use.[18] In analyzing the mechanisms of this impact, and the role of landownership patterns in explaining it, it is useful to look at two groups of counties: those that are already major coal-producing counties, and have been so for many years; and those that are currently more agricultural in their economic base, but which are currently facing coal development.

Most of the major coal counties are in the Central Appalachian region, and much of their land was removed from agricultural production long ago. The 1974 agricultural census, for example, lists only one farm in Mingo County, West Virginia. But to say that farming is no longer predominant in these counties is not to discount its significance. The small farm plot has provided important security for miners in times of coal bust, for the elderly and unemployed, or for those working in lower-paying jobs.

While the development of the coal industry took its toll on agriculture years ago in these counties, there has continued to be a loss of farmland even into recent years. This suggests that the last thread of independent economic security for residents in major coal counties is finally being eroded. Table 26 shows coal-producing counties in the sample in which loss of farmland between 1969 and 1974 has been the most dramatic. The average coal county lost almost 30 percent of its farmland in this period, double the rate in still agricultural counties. Only 18 percent of the land in these counties is now in agricultural use, about half the proportion in noncoal counties of the sample.

The contribution of landownership patterns to the decline of agriculture in these coal counties is suggested by correlations between corporate and absentee ownership of land, especially of minerals, and indicators of agricultural decline.

The greater the corporate control of mineral rights in these coal counties, the greater the loss of farms between 1969 and 1974. The correlation increases in strength when corporate control of both surface and mineral rights is combined into an Index of Resource Control.[19]

Among the coal counties, corporate ownership of the land is associated with lower agricultural use of land. Of forty-two major coal counties, only eleven (26 percent) had a high level of land in agricultural use. Of these counties, nine (82 percent) had a low level of corporate control.[20]

Table 26. Coal Counties with High Loss of Farmland, 1969–1974

	Percentage Loss of Farms	Percentage Loss of Farm Acres	Loss of Acres	Annual Coal Production
Knott, Ky.	79.6	67.6	9,174	4,321,000
Dickenson, Va.	60.8	49.4	10,282	5,299,000
Buchanan, Va.	59.8	62.1	16,382	15,804,000
Perry, Va.	57.6	48.1	5,350	7,473,000
Martin, Ky.	54.0	32.6	1,986	8,160,000
Logan, Ky.	51.2	49.2	2,289	8,612,000
Floyd, Ky.	45.5	42.2	13,821	4,562,000
Johnson, Ky.	45.1	40.3	20,667	3,810,000
Wise, Va.	44.1	40.2	8,757	12,290,000
Pike, Ky.	40.9	35.3	7,097	19,002,000
Letcher, Ky.	39.8	56.1	5,792	4,126,000
Lincoln, W.Va.	39.7	31.8	17,301	172,000[a]
Raleigh, W.Va.	39.0	34.5	16,157	6,828,000
Knox, Ky.	33.2	8.5	5,998	1,112,000
Breathitt, Ky.	31.8	35.8	30,451	6,373,000
Anderson, Tenn.	30.9	26.1	14,928	1,660,000

[a]Of these, Lincoln County is the only one without over 1 million tons of coal produced. However, land and minerals are tightly controlled by energy companies, and have been subject to heavy leasing.

Disappearing Farmland

The situation in Harlan County, Kentucky, provides a good example of what is happening to farming in the coal counties. In Harlan County, only 2 percent of the land is now used for farming, some 6,600 acres. Only forty-six farmers were listed in the 1974 agricultural census as farming this land. Thirty-eight of these had an annual income of less than $2,500, and twenty-two had an income of less than $1,000. Only fifteen farmed full-time. Twenty-five were at or near retirement age. Case studies from major coal counties document the problems farmers face in holding onto their land, and making a living from it. Little land is still available for agriculture use, and what there is may be threatened by the effects of mining. Unchecked strip-mining disturbs the land, fills creeks with silt that encourages flooding, and creates acid drainage that ruins the land it floods for future crops. Seventy-five percent of Cranks Creek in Harlan County is estimated to have been disturbed by strip-mining. What this means for local residents is that the creek is silted up, and most of the land below the strip job is ruined. Becky Simpson, a Cranks Creek resident, says, "Folks can't farm anymore, because the clay mud has washed over the soil; the land no longer absorbs water."[21]

On the fringes of the traditional coalfields, especially in southern Tennessee and northern Alabama, there are counties where agriculture has been the traditional economic base, but minerals are present and their exploitation is beginning to occur. In these counties, corporate and absentee ownership of minerals are coming into increasing conflict with local farmers' use of the surface land. A representative of the De Kalb County (Alabama) Soil Conservation Service says that the dramatic increase in strip-mining for coal over the past ten years has taken a great deal of farmland out of production in some areas of that county. In De Kalb County, farmers have reportedly banded together at times to buy land as a measure to prevent its purchase by absentee interests.

In the southern Tennessee counties that are now being exploited for their coal, there were several court decisions in the mid-1970s which backed the right of mineral owners to strip-mine land without the consent of the surface owners. In response, the state representative from one of the affected counties, White County, with the citizen's group Save Our Cumberland Mountains, pushed a bill through the state legislature in 1977 to force mineral owners to gain the consent of surface owners before mining. Though the law was challenged, it has recently been upheld by the Tennessee Supreme Court.

Strip-mining of land is the most obvious instance where coal development may act as a barrier to agricultural use of the land. Other effects of corporate and absentee control of land and minerals, especially the price spiral, may also have adverse effects on agriculture. Gary Kobylski, of the Walker County (Alabama) Soil Conservation Service estimates that the lowest selling price for farmland in that county is around $1,000 per acre, although some companies have offered farmers as much as $5,000 an acre. This price escalation occurs in a county where concentration of landownership by the coal industry has taken up to 20,000 acres of farmland out of crop production. A new regulation to preserve agricultural land by prohibiting mining of any land that has been planted in crops for five of the last ten years seems only to have encouraged speculation. Companies simply purchase the land and keep it out of production for five years.

In the agricultural counties of the sample there is a strong negative correlation between corporate and absentee control of mineral rights and the percentage of the county used for farming. This correlation is even stronger for the Index of Resource Control, combining surface and mineral ownership.[22]

Where there is a high degree of corporate ownership of land, and especially of mineral rights, there is also a high proportion of farmers who turn to other jobs to supplement their farm incomes. Since this land is taken out of the local market, either by price or by unwillingness to sell, and since the actual exploitation of coal under this land involves the destruction of the surface, farmers cannot expand their acreage to increase production efficiency, and new farmers cannot easily get a start in the occupation.[23]

The impact of coal development in these agricultural counties is only beginning: as more minerals are bought up, and as they begin to be exploited, agriculture will be more widely impacted. One may expect to find patterns developing in these counties that are now more clearly apparent in the "old coal" counties—a decrease in the use of land for farming, an aging farm population, a barrier to young people getting a start, an increasing pressure to turn to other jobs as a source of income.

It may be suggested that the move from an agricultural economy to a coal economy is not necessarily bad for the residents of the region. However, there is evidence that a decline in agriculture is associated with economic disadvantages for local residents.

LOSS OF FARMLAND: THE ECONOMIC IMPACTS

In the sample counties, the agricultural counties seem to be economically better off than the coal counties, despite the great wealth of natural resources in the latter. In the Virginia sample, the median family income in the coalfield counties is only 63 percent of the state average, while in the agricultural counties it is over 70 percent of the state average. The coalfield counties also have a higher proportion of families living at or below the poverty level than do the agricultural counties. The reasons become clear from case-study examples. Agriculture has in many cases provided a cushion against less stable sectors of the economy (whether coal, with its boom-and-bust cycles, or tourism). In Walker County, Alabama, agriculture is given credit by local authorities for carrying the county through the coal bust of the 1950s, when almost all the 7,000 jobs in the coal industry in the county were lost. Agriculture is still a significant sector of the economy in Walker County (employing 22 percent of the workforce, compared with the 24 percent employed in the coal industry). A balanced and diversified local economy, like Walker County's, has a greater chance of surviving economic hardships unscathed than the one-industry economy found in many other counties.

Other case studies illustrate the economic advantages of a significant agricultural base. De Kalb County (Alabama), a predominantly small-farm county with a well-distributed landownership pattern, had an unemployment rate of only 5.8 percent in 1979, compared with the state average of 7.1 percent. In Shelby County (Alabama), for all income indicators of economic health, the farm population was possibly better off than the nonfarm population.[24] Of the 1,960 farm adults recording income in Henderson County (North Carolina) in 1979, 59 percent or 1,166 had incomes in excess of $20,000. Another 26 percent made between $2,500 and $20,000.

The effects of agricultural decline can be seen in Grayson County, Virginia. In 1950, when agriculture was still a dominant part of the county's economy, and 44 percent of the workforce was employed in farming, the average weekly wage was 83 percent of the state average. By 1977, when only 16 percent of the county's workforce was employed in agriculture, and the county's economic base had changed to small-scale manufacturing, the average weekly wage was only 58 percent of the state average. The proposed federal and state developments that will lead to a tourist- and recreation-based economy in the county, with its low wages and seasonal employment, is unlikely to improve this ratio.

Even in counties where agriculture continues as a significant portion of the local economy, the impact of landownership patterns may be to make it less diversified and stable, less of an effective cushion against economic depression, than in the past. Production may become focused on crops that will yield a relatively large cash income on small areas of land. In western North Carolina this has meant ornamental shrubbery and Christmas trees; in parts of Alabama this has meant poultry; in other areas it means tobacco. Robert Thornton, the county extension agent in Walker County (Alabama), attributes the development of the broiler industry in that county to the lack of availability of land and the high price of land. Any of these limited (one-crop) farm enterprises are vulnerable to economic changes in ways that a diversified-food-crop agriculture may not be.

Several studies in other areas of the country have discovered a clear and direct relationship between small farms and a high level of social and economic development in small rural communities. The most important of these studies reported that as compared to a community surrounded by large farms, a small-farm community had twice as many businesses, 61 percent more retail trade, and three times as many household and building-supply purchases. It supported more people per dollar of agricultural production, had a better average standard of living, a much greater proportion of independent businessmen and white-collar workers, more and better schools, and twice as many civic organizations, churches, and means of community decision-making.[25] While drawing exact parallels between communities analyzed in the above study and rural Appalachian communities may be risky, such findings should prompt serious consideration of the positive effects of small-farm agriculture.

Other arguments also attest to the viability (even desirability) of a small-farm-based agriculture and would support whatever efforts are necessary to prevent the loss of small farms in Appalachia. The ecological argument suggests that the farming practices utilized on small farms are more ecologically sound than those on large farms. The efficiency argument maintains that the small farm can be just as efficient or more so than the large farm. Even the Ralston Purina Company, with long experience as a corporate farmer, admits that the family farmer "can meet and many times surpass the efficiency of large units that operate with hired management."[26] The political argument for small-farm-based agriculture suggests that political democracy is impossible without economic democracy and that the latter is enhanced by a diversified system of agriculture based on the widely dispersed ownership patterns typical of small-farm agriculture.[27]

While the economic and social advantages of small-farm agriculture are clear, policy strategies to promote it are rare. For example, the Wythe county (Virginia) Comprehensive Plan views agriculture as an important component in a diversified economy. Yet, while the plan seeks to "promote" industrialization, it seeks only to "protect" agriculture. For regional policymakers, the small farm has been largely ignored as "inefficient." The Appalachian Regional Commission has almost no programs directed toward small farmers. Nevertheless, the economic and social advantages of the small farm in the local economy must be recognized, as well as the other arguments in favor of the small farm.

Land is important historically and culturally to Appalachian people. It has been in the past, as Gladys Maynard of Martin County (Kentucky) puts it, "the people's survival kit." Economically, it has provided cash to counter the low wages and marginal employment often found in their rural communities, and it has offered some security against boom-and-bust industries. Appalachians have struggled to preserve their land, their values, and their lives as they know them. They are losing this struggle, in part owing to the nature of land ownership in the mountains today.

SIX

Homeless in the Mountains

> If you can't have homes and everything is choked to death, how is anything going to grow? We got the people, if we had something to build with we could go on.
>
> A lifelong resident of Mingo County, West Virginia

Housing in Appalachia has long been recognized as a national disgrace. In 1970, in the region as a whole, one out of every five homes was considered substandard. In Central Appalachia, the figure rose to one in every three homes. Of the seventy-two rural counties in this study, the average county had 30 percent of all homes lacking some plumbing, 13 percent considered overcrowded, and almost 60 percent built before 1950. For people living in the region, these statistics are made worse by the paradox that some of the worst housing conditions lie amid the greatest wealth. In the heart of the Appalachian coalfields, houses are among the oldest and most overcrowded. In the recreation and tourist counties, substandard, locally owned dwellings stand side by side with modern, absentee-owned second homes. Throughout the region, mobile-home parks along the roadways and riverbanks have been the principal solution to the lack of adequate housing.

A multitude of reasons have long been given for the persistence of Appalachia's housing crisis. Mountainous terrain, lack of water, sewage, and other services, shortage of capital, and frequent flooding are among them. In recent years, growing reference has been made to another problem: the barriers that landownership patterns pose to decent housing. In West Virginia, the Governor's Housing Advisory Commission reported, "A related problem in coal-mining areas of the state is that most of the developable land is owned or controlled by natural resource companies. The speculative value of the property makes it nearly impossible for builders to purchase a permit that permits development of low and moderate housing."[1] To this the 1980 President's Coal Commission added, "The land shortage in Appalachia is, in part, attributable to coal companies, railroads, and other corporations owning much of the coal-rich acreage. With future plans to mine their holdings, companies prevent their unimproved properties from being developed."[2]

While the problem of landownership's impact on housing is recognized, its extent and complexity has lacked systematic study. The President's Coal Commission stopped short of so doing, pointing out that "statistics for land ownership are often buried in inacessible or untraceable county records."[3] However, data obtained in this study allow for the first time an in-depth exploration of the role of landownership patterns in Appalachia's housing crisis.

Theoretically, Appalachia has abundant land for its housing needs. In 1970 in the average rural county in the sample, there was only one house per every thirty-five acres of land. But much of this abundant land lies empty and inaccessible to the region's people. Over one-half of it is owned by absentee owners, corporations, government agencies, and large holders who value it for its mineral or timber resources, for its recreation potential, or for its speculative value—not for meeting local housing needs. Still other land is uninhabitable, or is used for farmland, roads, schools, industry—the result being a land shortage in the midst of a land-rich region. Interviews in numerous counties document the pattern: land for housing is often simply unavailable for purchase.

LACK OF AVAILABLE LAND

From the tightly packed valleys of West Virginia to the open plateau and rolling hills of Alabama, people say that land for housing is just not available to them. In Walker County, Alabama, a representative of Farmers' Home Administration said: "The land situation is this: land is tightly held by coal and timber concerns. Very little turnover of land occurs, the vast majority of turnover being among family members."

In Harlan County, Kentucky, the housing market is going from bad to worse. In 1978 there were 13,413 units, 53 percent of them substandard. "There is no space to build because companies own so much land, and the companies won't sell a piece of land as big as a desk," says a local miner. In Martin County, Kentucky, there is a desperate need for more houses. In May 1977 there was a vacancy rate of only 0.3 percent in Martin County. Thirty-one percent of the county's occupied houses were classified as substandard. A housing plan prepared by the Big Sandy Development District notes the role of corporate owners in adding to the pressure for housing in the county. "The coal companies are directly responsible for many recent events in the housing market, and own up to 50 percent of the land in Martin County. Many homes have been bought in the hollows at fairly high prices, and families displaced then joined the incoming workers in the search for housing in Inez and Warfield." The editor of the local newspaper, Homer Marcum, puts even more strongly the connection between corporate landholding and the county's housing shortage: "The average individual who must work for a living doesn't stand a chance of getting any land from them (the companies); he is simply left out of consideration."

In northeastern Tennessee coal counties, there is a similar pressing shortage of housing for local residents. In Campbell County, there was a shortfall of 6,269 units in the 1980 housing supply, 52 percent of the total number of households now in the county. In neighboring Claiborne County, a nonprofit citizens' group asked American Association, Inc., a major landholding company there, to make available a small tract of land to build sorely needed houses for the local population. Although the company owned tens of thousands of acres of land, it refused to provide any land to meet housing needs.

Ownership of mineral rights extends the control gained from concentrated ownership of land, and further restricts the possibility of housing development. Throughout the coalfields there is extensive separation of mineral from surface ownership. Residents who own surface land without the underlying mineral rights are subject to many uncertainties: companies may show up to strip-mine the land at any time; conflicts may develop over title. Severed ownership of mineral rights also affects home building, through restricting the availability of loans. As one bank officer in Dayton, Tennessee, explained, lack of mineral rights acts as a "cloud" on the title, and title companies will not insure it. Without title insurance, lending institutions—including Housing and Urban Development and Farmers' Home Administration—will not make loans, and neither first nor second mortgages are available to these property owners. A Tennessee resident, Raymond Weaver of Sale Creek, outside Chattanooga, can attest to such policies. A post office employee, he can show papers from more than five lending institutions that turned down his application for money to renovate his home. The rejections were made because he does not own the mineral rights beneath his forty-six-acre farm.

The scarcity of land for housing created by concentration of ownership in large blocks also drives up the prices of what land is available for sale. The consequent inflated price for land affects residents in coal counties and recreation counties alike. In coal counties, local residents must compete with energy companies and land speculators; in recreation areas they compete with second-home buyers and resort developers. The effect is the same; to place even small tracts of land out of the price reach of most local residents, especially low-income and blue-collar families.

In Walker County, Alabama, the agricultural extension agent says, "The price of land is now based on the value of the underlying minerals, whether it is to be used for agricultural, housing, or mining purposes." In the rural part of his county, homesites now range from $2,400 to $3,000; while near Jasper such a lot would sell for $5,000-$7,000. In Walker County, the per capita income in 1974 was only $3,345.

In other coalfield areas, scarce land fetches similarly inflated prices: In Martin County, Kentucky, where the demand for housing is so high that only 0.3 percent of housing is unoccupied, the price of housing has more than doubled in the last five years, according to the director of the County Housing Agency. County Planner Larry Smith says corporate purchases of coal lands at unusually high prices have eaten into the county's stock of residential land and helped to drive up land values. In neighboring Pike County, Kentucky, almost all of the developable land in Elkhorn City is owned by the Elkhorn City Land Company. Its 1,405 acres are assessed for taxation purposes at $36 per acre. Each year the company sells two or three lots for housing, each 50 feet by 100 feet. The price is $20,000-$25,000 each. Very high land prices are found in Harlan County, Kentucky, too. The local development district has found two sites that it would like to develop for housing. A six-and-one-half acre tract is for sale by an architect at $500,000; an eight-acre tract is for sale by a Lexington physician for $250,000.

Lack of housing land available in coal counties affects neighboring counties. A young lawyer in Lincoln County, West Virginia, says: "Low- and middle-class families can't afford housing. One reason for the shortage is that people from Logan County coalfields have to live in Lincoln County. Logan coal companies own surface rights also, so people can't get housing there." The consequent pressure on housing in counties adjoining coalfields drives up the prices there as well.

While in the coalfields, local residents must compete with the prices energy companies can afford to pay for land, in the recreation areas residents face a similar price pressure resulting from second-home buyers and resort developers. In some of these counties, land availability for housing is already restricted by U.S. Forest Service ownership. Competition for the remaining land is heightened by urban dwellers with incomes far above those of most local residents, who pay prices for "a place in the mountains" that few local residents can afford.

Watauga County, North Carolina, illustrates the problems facing residents in recreation counties. In 1960, 5,554 housing units existed in the county, including 727 second homes. By 1970, 3,000 new homes had been added to the housing stock. Over 1,000 of these—more than a third—were second homes. According to the housing census, 21.2 percent of the houses in the county in 1970 were "seasonal and migratory"—likely resort and second homes intended for occasional occupancy.

Many more possible housing sites in Watauga County have been subdivided in recent years. Of the 129 subdivisions identified, 9 were recorded in the 1950s, 40 in the 1960s, and 80 in the 1970s. This reflects the impact of the last two decades' resort and recreational development, as well as population increases. But land subdivided for housing does not necessarily result in more houses available to the local population. Of the 10,000 platted lots recorded in the subdivision inventory, only 16 percent had houses built on them. Since many second-home lots are sold without any initial intent of construction, the county experiences the negative impacts of inflated land prices without the accompanying benefit of construction employment or additional tax revenues.

Speculation and subdivision of land have been major factors in driving up land values in Watauga County, as in others affected by recreation development. The tax base in Watauga County, reflecting these higher land prices, increased over 300 percent between 1961 and 1974. The cost of housing more than doubled, increasing far faster than the wage rate in the county. As a result, many local families have little hope of owning their own home.

In counties like Watauga, the housing pattern reflects a dual standard. Second homes in resort communities are often of higher quality and receive more services than the scattered, rural, often substandard homes inhabited by local people. The latter feel they subsidize with county revenues the second-home development, while at the same time having to bear higher land and housing costs. They resent the paradox.

Similar patterns to Watauga County's are found in other counties affected by second home and tourist development. In Cumberland County,

Tennessee, for example, subdivisions have sprung up as land values have increased. Ten years ago there were only two subivisions and 10,000 parcels of land in the county. Now, according to Martha Oaks, the county tax assessor, there are numerous subdivisions and 45,000 parcels of land. Housing is available in Cumberland County—if one can afford to buy. In the last fifteen years the price of land has risen from an average $100 an acre to $1,000 an acre. Land speculation and increased demand from recreation and residential development have served to place prices out of reach of low and middle income families.

If the relationship between landownership and housing problems revealed in these case studies is a general one, we might expect to find significant correlations between landownership patterns discovered in the eighty-county survey and such housing indicators as overcrowding and age of housing stock.[4] To avoid skewing by urban patterns, the analysis will be applied only to the seventy-two counties of the sample which are "rural" (i.e., more than 50 percent of the population live in rural areas.)

One indicator of housing shortage is the amount of overcrowding, measured by the number of homes with more than 1.01 persons per room. From the case study data, it might be expected that overcrowding will be correlated with degree of unavailability of land, connected with high corporate or absentee ownership. On average, a county in the sample has 12.4 percent of its housing units overcrowded (with more than 1.01 persons per room), compared to the national average of only 8 percent overcrowded housing.

In these seventy-two counties, the general relationship holds: the greater the proportion of corporate land, or of absentee-owned land, the greater the proportion of housing units that are overcrowded. Of twenty-six counties with a higher-than-average degree of corporate ownership, twenty-five (96 percent) also had a higher-than-average proportion of overcrowded housing. Of forty-six counties with a low degree of corporately controlled land, only twenty-three (50 percent) had above-average proportion of overcrowded housing.[5] A similar pattern holds for absentee ownership: of forty-seven counties with a higher-than-average degree of absentee ownership, thirty-five (74 percent) had higher-than-average overcrowded housing. This compares with the twenty-five counties that had low absentee ownership, of which 52 percent had higher-than-average overcrowded housing.[6]

A further measure of unavailability of land for housing can be compiled by combining the degree of government ownership in a county with the degrees of corporate and absentee ownership. This measure of unavailability also correlates with the proportion of overcrowded housing.[7] Of counties with high levels of overcrowded homes, 71 percent also have a high proportion of "unavailable" land. (See Table 27.)

One might respond that the coal counties, with their denser populations, are skewing the relationships here. However, even within the category of coal counties, the relationship holds: 63 percent of the coal counties with high levels of overcrowded housing have high proportions of "unavailable" land. Eighty percent of coal counties with less crowded housing show greater local control of land.[8]

Table 27. Correlation of Unavailable Land with Overcrowded Housing in Seventy-two Rural Counties

Percentage of Unavailable Land	Percentage of Overcrowded Housing		Total
	Low (less than 10%)	High (10% or more)	
Low (less than 40%)	20[a] (58.8%)[b] (83.3%)[c]	14 (41.2%) (29.2%)	34 47.2%
High (40% or more)	4 (10.5%) (16.7%)	34 (89.5%) (70.8%)	38 52.8%
Total	24 33.3%	48 66.7%	72 100.0%

Pearson's R = .411 at the .0003 level of probability.
Note: All land owned by corporations, government, and absentee owners, i.e., not by local individuals, is defined as "unavailable" for local housing. "Overcrowded" is defined as more than 1.01 persons per room according to the 1970 Census of Housing.
[a] Number of counties.
[b] Percentage unavailable land (row).
[c] Percentage overcrowded housing (column).

Within the noncoal counties, as a group, the connection between unavailability of land and overcrowding also holds. Of the noncoal counties with a high degree of unavailability of land, 75 percent have a high level of overcrowded housing. By contrast, of the noncoal counties where land is more likely to be available (because less is held by corporate, absentee, or government owners), only 7 percent had a high level of overcrowded housing.[9]

While the effects of unavailability of land on housing may be the same in coal and noncoal counties, it is likely that the mechanisms at work differ. In noncoal counties, the impact of corporate landholding on overcrowded housing is not found to be statistically significant. However, the relationship between absentee ownership and this housing indicator is strong. All of the noncoal counties that have a high degree of overcrowded housing also have a high rate of absentee ownership. In these noncoal counties also, the degree of government or public, nonprofit ownership is related to the degree of overcrowded housing. As one would expect from the case-study findings, the aggregate data confirm that in noncoal areas it is primarily the absentee ownership of second homes and recreation developments or government and private, nonprofit landownership that makes land unavailable to local residents.[10]

In the coal counties, on the other hand, the impacts of energy development on housing are the same whether the energy land is owned by corporations or absentee individuals.[11] In both cases, the housing shortage is exacerbated by landownership patterns that help to keep land out of the housing market.

If, as the data presented so far suggest, landownership patterns act as a significant barrier to new housing development in Appalachia, one may expect a further correlation between concentration of landownership and age of housing stock. Such a statistical relationship is indeed found here. In the seventy-two-county sample, 59 percent of the homes were built before 1950 (this compares with a national average of 48 percent). In the coalfield counties this proportion rises to 64 percent (compared with 53 percent in noncoal counties). One might expect many of these older houses in the coalfields to be in the coal camps, built before the 1950s slump in the coal market, when coal companies were principal housing providers for the miners and their families.

Within the coalfield counties themselves, there is a significant relationship between older housing and the degree of corporate control of land. Such a correlation suggests that where corporate owners hold large amounts of land, little becomes available for new housing to be built. In Mingo County, West Virginia, for example, where corporate ownership of land and minerals equals 180 percent of the county's surface acreage, 83.2 percent of the housing was built before 1950. It should be noted that, in addition, almost a third of the county's housing stock was torn down between 1950 and 1970, not to be replaced. Altogether, in the four West Virginia counties that lie at the heart of corporate control of the southern West Virginia coalfields—Mingo, Logan, Raleigh, and McDowell—71 percent of the housing stock in 1970 was over 20 years old.

Throughout the coalfield counties of the sample, the relationship holds: the greater the corporate control of land, the greater the proportion of older homes in the county. While the general relationship is not a strong one, it should be noted that outside the coalfields, no statistical relationship was found between corporate control of land and age of housing.[12]

When mineral rights are considered within the coalfield counties, the relationship increases in strength, helping to confirm the case-study findings that severed mineral rights act as an additional obstacle to home building. They place a "cloud" on title, making loans difficult to secure. In general, the statistical correlations suggest that the greater the control of mineral rights (apart from consideration of the surface), by corporations, government, and absentee individuals, the greater the degree of housing built before 1950.[13]

When concentrations of control of both surface and minerals are combined, these relationships still hold: in the coal counties, the greater the extent of resource ownership by corporations, government, and absentee owners, the older the housing supply.[14] In general, the data support the hypothesis that it is landownership patterns which serve as a barrier to housing development in Appalachia. This argument counters a conventional explanation of Appalachia's housing shortages that the shortage of flatland is the primary barrier to housing development. The conventional argument about the source of Appalachia's housing shortages—the terrain argument—makes assumptions about the availability of land that do not withstand careful scrutiny.

First, the flatland argument assumes that if more flatland existed in the region (or if more land could be artificially flattened), it would be made available for housing. The concentration of ownership in the hands of absentee and corporate owners documented in this study suggests otherwise. These owners value the land for its mineral, timber, or other resources, not for its potential to house local people. As Ernest Chaney, of the Pikeville, Kentucky, Housing Authority, says, "One hundred years from now, the coal companies are going to be going for the coal under the flatland." As long as the land has other value for its owners, it is not likely to become available for housing at reasonable prices.

The terrain argument also assumes that housing shortages are found only in the mountainous areas. In fact, the indicator of housing shortage used above—the degree of overcrowding—is found in counties with all types of terrain. From the relatively flat land of Walker County, Alabama, to the steep hillsides of Harlan County, Kentucky, chronic housing problems exist.

The Tennessee counties of this study serve as a graphic illustration of this point. If lack of flatland were the key to housing shortages, one would expect to find significant overcrowding only in the more mountainous counties of the study—mainly Campbell and Scott counties. In fact, higher-than-average levels of overcrowded housing are found in these two counties, but also in the plateau counties of Fentress, Bledsoe, and Sequatchie, where terrain is not an intrinsic barrier to home construction. In Tennessee, a better explanation of the overcrowding in these counties is found in the concentration of landownership. The index used here is the percentage of surface and mineral acres owned by the top five landowners in a county. High concentration of ownership seems to be closely correlated with higher than average levels of overcrowding. The converse is also true: counties with low concentration of ownership also have low levels of overcrowding. (See Table 28.)

Finally, the terrain argument for Appalachia's housing shortages does not stand up to historical scrutiny. In the past, more housing units existed in many parts of Appalachia than exist today. They were in coal camps, provided by corporate landowners for the families of the many miners who were needed to run the deep mines that thrived then. While the quality of this coal-camp housing may have left a lot to be desired, the fact remains that housing sites were there, which are not available today. What has changed is not the terrain but the policies of the corporate landowners.

An example of how corporate controllers of land have changed their policies and taken land out of the housing supply is in the Clear Fork Valley of Campbell and Claiborne counties, Tennessee:

> Once a prosperous mining valley much of the valley's land now lies offlimits to its residents. Whole coal camps, like Westbourne, have

Table 28. Impact of Control of Surface and Mineral Rights on Overcrowded Housing in Fourteen Tennessee Counties

Percentage of Surface + Mineral Rights Controlled by Top Five Owners in County**	Percentage of Houses with More than 1.01 Persons Per Room*	
	High	Low
High	Campbell Scott Fentress Bledsoe Sequatchie	Van Buren
Low	Cumberland Marion	Anderson Morgan White Rhea Hamilton

*1970 Census of Housing. High or above average is greater than 13% of the houses in the county.

**Total surface and mineral acre ownership expressed as percentage of total county surface. High or above average is greater than 33.3% of the county surface.

simply disappeared. Since the 1880's, the valley has been dominated by a single large corporate owner—American Association Ltd., a British company. It leased its coal to smaller companies to mine, and these in turn built the coal camps for their miners' families. In 1950 there were 10 large underground mines in the small Claiborne County section of the valley alone, employing some 1,400 men. The valley had one major community, Clairfield, and many surrounding coal camps. During the 1950's, however, as in the rest of Central Appalachia, the mines began to close, and the valley's people joined the migration to the cities of the North in their search for jobs. As the mines closed, American Association took possession of the coal-camp homes. It had no interest in maintaining the homes, or the communities. The company manager went on record as saying, "The people would be better off, we would be better off, if they would be off our land."

More than two-thirds of the company houses were torn down and not replaced between 1962 and 1972. The company made it clear to residents that they were not welcome. Leases, if granted at all, were for only 30 day periods. Notices were posted at the stores, mines and post office, saying, "No specified reason is needed if the owner desires to have the house vacant.... No one is obligated to remain in a house. If he is unhappy about his surroundings he is free to move immediately."

American Association accepted no responsibility for the communities it was destroying. In an interview with a British Broadcasting Company team in 1974, the company manager in Middlesboro, Kentucky, Alvarado E. Funk, was asked:

BBC: Don't you have a sort of moral responsibility to maintain the people who wish to stay in that area, and who could have been working their fingers off to keep them in a reasonable condition of living?

FUNK: No, sir, these people don't work for us and never have worked for us—they're just people.

BBC: But they're living on your land, aren't they?

FUNK: We don't have any responsibility for them ...

BBC: You mean they get in the way of strip-mining operations?

FUNK: Well, I don't say they get in the way, but they just don't add anything to the assets of the company.[15]

Throughout the coal camps a similar policy shift occurred: the industry collapsed, people left the region, the houses were torn down. Now, the coal industry is booming again, but housing sites for the returning people are not available. Buildable land remains vacant as corporate owners refuse to make available land that housed previous generations of miners.

This pattern can be substantiated by comparing housing units in 1950 with those in 1970 in major coal areas. Altogether, in the twelve eastern Kentucky counties of the survey, there were 8,000 fewer housing units in 1970 than there had been in 1950. In Harlan County, Kentucky, where 75 percent of the land sampled in the survey is corporately held, there were 16,782 housing units in 1950; by 1970 there were only 12,446, a decline of 26 percent. But there are said to be no housing sites available in the region.

Similarly, in West Virginia in the four southern coalfield counties in the survey—Mingo, Logan, McDowell, and Raleigh—there were 12,579 more housing units in 1950 than in 1970. In McDowell County alone, the number of housing units declined in this period by almost a third. It is often said that these counties have the most rugged terrain, and that this is the cause of the housing shortage. The prior existence of more housing units in these counties refutes this argument. A more plausible explanation of the housing shortages there is that these counties are also the most tightly controlled by corporations; in this four-county area, almost 90 percent of the land sampled is corporately held, amounting to over two-thirds of the total surface.

Our data suggest that terrain is not so much of a barrier to housing development as are the policies of corporate landlords, once the major providers of land for housing in the coalfields. Much of their land suitable for housing development now lies empty.

The analysis presented so far suggests that landownership must be considered as a major factor contributing to housing shortages in Appalachia. The ownership patterns found in this study keep land unavailable to the housing market, and/or out of reach of low- and middle-income buyers. However, the importance of landownership in directly affecting housing shortages should not detract from the contribution of other factors. These other barriers to housing development in the region, often acknowledged in other studies, include lack of financing, lack of suitable infrastructure (notably water and sewage), the occurrence of repeated flooding, and the dearth of a construction industry. However, while giving due weight to these other factors, it is important to recognize that they too are affected by landownership and use patterns. Landownership has indirect effects as well as direct effects on the region's housing problems.

OTHER BARRIERS TO HOUSING DEVELOPMENT

Throughout the case-study interviews, local residents report the difficulties of obtaining adequate loans to finance land purchase, the building of new homes, and improvement of old houses. In part, these difficulties reflect current national financing problems—high interest rates and a tight money supply. However, these contemporary national problems are not new in Appalachia, where they are compounded by other problems peculiar to the region. And it is these particular features of financing difficulties that are influenced by the region's landownership patterns. The factors involved are demonstrated in both private-sector and public-housing financing.

It is ironic that many of the reports of tight financing for housing come from the coalfields, where vast amounts of wealth are now being produced from the region's natural resources. Enough capital is produced in these counties to develop local housing. Indeed, according to census data, banks and other financial institutions in the average coal county in this survey had some 56 percent more money on deposit than the average noncoal county. Rather, what these counties lack is the reinvestment of that wealth in the long-term improvement of the community.

The "time" deposits in local banks provide the major pool of lending capital, whether for economic or community development. In the average coalfield county of the sample, time deposits amount to only 64 percent of total bank deposits in the county, compared with 71 percent in noncoal counties. Some coalfield counties fare even worse. Harlan County, Kentucky, for example, where housing is especially bad, has only 24 percent of its bank deposits in "time" deposits. Capital flows out of the region for investment elsewhere, rather than becoming available for local development.[16]

In these coalfield counties, a statistically significant relationship exists such that the higher the degree of absentee ownership the lower the proportion of local bank assets in "time" deposits. While the relationship is not very strong, it should be noted that outside of the coalfields no statistical relationship was found.[17] The coalfield pattern suggests that absentee ownership of resources actually detracts from the possibilities of local development of housing, by restricting the availability of local private financing.

Lack of locally controlled capital leads to a lack of home-finance institutions. In the rural Appalachian region, savings and loan associations, nationally the principal source of home mortgage money, are few and far between. Even where lending institutions do exist, their policies may serve to exclude or restrict access of rural and poor people to what financing is available in a county. In Harlan County, for example local banks have required a 30 percent down payment on a home (during periods when the average down payment required nationally was 10 percent) and they required a shorter payback period (ten to fifteen years).

In other case studies, it appears that rural parts of the county do not gain as much in loan finances as the wealthier urban areas. In Scott County, Virginia, the Estivill magisterial district, containing the towns of Gate City and Weber City, has 40 percent of the county's population. Yet of the only two banks operating in the county, Virginia National lent over three times as much money in the Estivill district as in the rest of the county combined, and Bank of Virginia lent six times as much money there between 1975 and 1977. The Estivill district is considerably wealthier than the rest of the county, with only 21 percent of its population below the poverty line, compared with 35 percent in the rest of the county.[18] In Hamilton County, Tennessee, case-study interviews suggest that some banks discriminate against the county's rural population. Owners of property in the expensive Signal Mountain neighborhood seem to have had little difficulty obtaining credit from local banks, despite not holding the mineral rights under their land, while people in the more isolated rural parts of the county, such as Flat Top Mountain, Sale Creek, and Montlake, have been told that their lack of mineral title is a major obstacle to obtaining loans for building or renovating housing. Private-sector financing for homes seems to be fraught with difficulties for many Appalachian residents. At least some of these difficulties have a connection with the patterns of landownership and land use found in the region.

It is in part to compensate for the deficiencies of private-sector financing that programs such as the Farmers' Home Administration (FmHA) and HUD exist. Yet these programs, too, have failed to remedy the problem of Appalachian housing shortages. In Harlan County, Kentucky, where available housing falls far short of the population's needs, and private financing is hard to obtain, there was not one FmHA loan for a new home in 1979. Our study suggests several reasons why FmHA and HUD programs may fail in Appalachia:

1. They presuppose that land on which to build housing is available. In fact, such housing sites are extremely difficult to come by in many parts of the region.
2. They demand clear and "unclouded" title to the land, which often is not available, at least in the coalfields, where mineral rights are often severed from surface ownership.

3. The inflated prices produced by housing shortages may deplete the amount of funds in a particular area. In Walker County, Alabama, for example, FmHA in 1978 "spent it faster than they could get it" and, in the first month of the highest financed quarter of the year, spent all their allotment for that quarter. FmHA officials say that this is largely due to the extremely high price of land for homesites.
4. FmHA and HUD restrictions on physical site requirements severely limit their funding availability in some areas. In Harlan County, Kentucky, for example, the Harlan Housing and Urban Development Agency has had difficulty in getting site approval from HUD evaluators because there is no fire protection, police, city water or sewage, ambulance service, or shopping center. One of the local agency staff describes this as "a basic contradiction between federal regulations and the reality of life here."

Insofar as the lack of local services is a barrier to federal financing, the low tax base of these counties is partly to blame. And, as Chapter 3 has shown, this is associated with land ownership patterns in the region. Insofar as physical features such as flood-plain restrictions and water supply are to blame, these too are affected by landownership and land-use patterns.

The inadequate development of a service infrastructure—roads, water, sewage systems—has often been blamed for Appalachia's housing problems. Certainly these services are lacking. In the eighty counties of this survey, over 90 percent of rural homes lacked sewage service. And nearly 43 percent of the homes in the average county of the sample lacked some plumbing. Roads in rural areas are generally poor, ill-paved, or not paved at all. In the coalfields, coal hauling, much of it in overweight trucks, has resulted in severe deterioration of secondary roads, which were neither designed nor built for such traffic.

Several factors play a part in the infrastructure deficiencies of Appalachia. Some of them, in turn, are related to landownership and land use. Ownership patterns of large blocks of land that are unavailable for housing combine with mountain terrain to make delivery of water and sewage systems expensive. Houses are scattered in isolated pockets, or strung out for miles along narrow valleys. Underlying and compounding these difficulties of service delivery is the lack of adequate tax revenues in these counties with which to provide service to residents. As Chapter 3 details, property-tax structures in the region are regressive and deficient, and do not generate enough capital for local services. Taxes per acre are lower in the counties where landownership is most concentrated. It is these same counties where our analysis above has suggested there are already the most barriers to overcome in order to develop housing. The tax structure only compounds their inherent problems.

Appalachian counties' inability to provide sewage services to rural residents is symptomatic of the problem. In the eighty-county sample, the average expenditure per capita per year by county governments on sewage services was $.83, amounting to less than a half of one percent of total county expenditures. In fact, in the 1977 Census of Governments, in only seven of the eighty counties were any county sewage expenditures reported.

The lack of available services may render scarce land that does become available for housing unusable, or unfinanceable. In Harlan County, two blocks of land that might be developed for housing remain empty. One, a ninety-nine-acre tract owned by the Chamber of Commerce, has gone undeveloped for nine years because there is no access bridge across a river. Another eighty-three acres, donated by the Eastover Mining Company for residential development, remains empty because of lack of water services. One local housing-agency staffer maintains that so long as HUD holds to its floodplain and sewage regulations, 92 percent of Harlan County will remain ineligible for HUD monies. Many other Appalachian counties, especially in the coalfields, are under a similar disability.

Even where water and sewer services are provided, they may discriminate against local residents. In the resort counties in particular, the interviews suggest that these services may be more available to absentee, second-home buyers and resort developers than to local residents. Local people believe that the developers have more political influence, and use it to get services delivered. For example, in Campbell County, Tennessee, several families had lived in the Shady Cove area for years without city water. All their attempts to get water lines extended to them had been in vain. In 1978 a developer constructed an exclusive vacation-home subdivision about one mile from Shady Cove. The water line was extended to the new subdivision, bypassing the Shady Cove residents and arousing hostilities between local residents and the developers.

The example of Shady Cove is not an isolated one. In the nineteen tourism and recreation counties of the eighty-county sample, we found a strong correlation such that the greater the proportion of absentee landownership (likely resort developments), the greater the percentage of rural homes with sewage services. This correlation was not found for any other type of county.[19]

Another argument given both for lack of suitable land, and for difficulties in financing, is that many available housing sites are in the floodplain. Certainly, flooding has taken its toll on housing in Appalachia, particularly in the Central region. The April 1977 flood destroyed 600 homes in the Tug Valley area and 600 more in Harlan County. In the two areas together, over 5,000 more homes were damaged. Smaller floods persistently rack Central Appalachian valleys.

It is important to recognize, however, that the causes of flooding are at least partially related to landownership and land-use patterns. Historically, corporate ownership has been associated with the higher areas away from the floodplain. This pattern emerged partly because the valuable, cultivable land along the river bottoms was more difficult for the coal companies to obtain from local farmers than were the hillsides, and partly due to the geology of the region that made coal seams on the mountainsides more accessible for mining. Regardless of the cause, the areas along the river bottoms traditionally have been left for housing and small farms.

With the advent of strip mining and other destructive land uses in the mountainsides, however, the flooding in the bottomland has become more frequent and more destructive. A growing number of studies now establish the link between strip-mining practices and flooding (see Chapter 7). The combination of the ownership and the use pattern is serious for housing: while higher lands are owned by the corporate and absentee holders who use it for energy extraction, their use of that land limits the possibility of housing in the valleys.

It is little wonder, given these various obstacles to housing in Appalachia, that many parts of the region also lack a building industry. Traditionally, as has been seen, the coal industry was the major housing supplier. While the industry no longer is building, few new opportunities have emerged for developing and marketing affordable homes. Even where housing projects do develop, according to housing experts, local builders cannot sustain their business, owing to uncertainties about when land will become available for the next project.

HOUSING ALTERNATIVES

Unable to buy land or their own homes, many Appalachian residents have only two options available. Both fall far short of being acceptable alternatives. Throughout the coalfields, many rented homes remain, despite the destruction of so many coal camp houses in the 1950s and 1960s, and in coal counties and noncoal counties alike, mobile homes increasingly dominate the housing scene.

In the coal counties, the extent of company housing is suggested by the strong correlation between degree of corporate ownership of land in the county and extent of tenant dwellings.[20] In the average coal county of this sample, 31 percent of the housing units in 1970 were rental units. In noncoal counties only 25 percent were in this category. In some of the Central Appalachian coal counties, the figure climbs even higher—to almost 40 percent rental units in Harlan, Bell, and Breathitt counties, Kentucky, and in McDowell, Mingo, and Logan counties, West Virginia.

This relationship also confirms the extent to which corporate ownership of land acts as a barrier to people building or obtaining their own homes. For many Appalachian people, coal camp life is not a bygone era. Facing no alternative, people remain, often dependent upon the will and wishes of the company landlord. In staying, they face insecurities of tenure, dilapidated housing, and fear of the company's power.

An example of this state of affairs is in Logan County, West Virginia, along Rum Creek, where the land and housing is owned by the Dingess Rum Coal Company. In Logan County, hundreds of coal company homes were destroyed during the coal slump of the 1950s and 1960s. Now, even though the housing crisis is desperate, the land where those houses stood lies vacant, and the companies refuse to sell. The coal industry is expanding now and houses are needed for miners, but Dingess Rum continues to tear down livable housing as tenants die or move out. Along Rum Creek, residents have heard that the company now plans to tear down what housing remains. Richard Cooper, a United Mine Workers safety inspector who lives in a company house, says that Dingess Rum officials recently got tenants to sign a form agreeing to vacate their homes within ten days if the company asks them to. "We used to have a thirty-day-notice period before they could put you out. They just lowered that to ten days. You have no choice. You sign or you're gone."

Cooper knows the policies of the company well. He grew up on Rum Creek, where his father rented a company house. Now Cooper, his wife Phyllis, and their three children live in a Dingess Rum house at Yolyn that is at least fifty years old. The Coopers pay $89 a month in rent for the house, which sags with age. The roof has a gaping hole in it and water sprays from broken pipes under the house. But the Coopers do little to improve the house, because the rent will go up if they do more. The Coopers would like to buy land on Rum Creek for a house. But the company flatly refuses to sell. "I could go up and offer $100,000 for this house and they'd laugh in my face, even if I had it in $100 bills."

In Rum Creek, and throughout the coalfields, tenancy combines with the lack of alternatives in both housing jobs to place power in the hands of the landlords. An example is seen in the small community of Braden's Flats in the upper east Tennessee coalfields, where most residents are tenants of the Coal Creek Mining and Manufacturing Company. In 1979, the company leased land for strip mining within a few hundred feet of several families' houses, and applied for permission to close the county road into the community in order to extract its underlying coal. In what might, in other situations, have been a controversial matter, all the affected residents of Braden's Flats gave permission for blasting operations, and indicated little opposition to the disruption of their road. Their fear of the "company" is all too common in coal camp communities.

For those not dependent upon the coal camp for their housing, the other option is often the mobile home. In parts of Appalachia, the trailer park appears to have replaced the company town. Again in Logan County, a resident says: "It seems that the general policy of Dingess Rum is to make their housing as unbearable as possible in order to coax county residents into trailer camps. Today, Dingess Rum makes as much renting families plots of land on which to place a trailer as they used to make renting housing. And, they pay less taxes, because the land is considered idle for tax purposes." Case-study material indicates the rise of new mobile homes in Appalachia to be staggering. In seven coal-producing counties of southwest Virginia, a record number of occupancy permits was issued between 1 January and 30 June 1979. Of the 1,335 permits, 1,012 or 76 percent were for mobile homes. In Wise County, Virginia, mobile homes accounted for over 70 percent of the new housing units between 1970 and 1976. In Pike County, Kentucky, mobile homes represented 98 percent of new housing units between 1970 and 1977.

For many, the mobile home is an easy way to bypass the obstacles to housing that have been identified in this chapter. Unable to buy land on which to build, a family can squeeze a trailer onto a small plot of family land, or place it in a trailer park. Unable to get financing for a house, a family can make the small down payment on a trailer with minimal credit problems. Unable to get the services needed for home building—sewers, water, roads—a family can move into a trailer park where the services are immediately available.

Yet a number of problems arise. Trailers are essentially a short-term solution to a long-term problem. Their life expectancy is much less than that of conventional housing. The housing crisis will still exist in ten or twenty years when the trailers are no longer inhabitable.

The crowded nature of trailer parks, and additions of trailers onto small plots of family-held land, radically changes the rural nature of many Appalachian counties. In Pike County, Kentucky, according to a survey conducted for the Pike County government, in 1978, the phenomenal growth in mobile homes has resulted in overcrowding of creeks and hollows, and virtual elimination of farming. There were 828 trailers in the county in 1970; 6,389 by 1977. In Wise County, Virginia, 74 percent of the population lives in the 2 percent of land area that is classified as "urban and built up." The population density of this area is 4,035 persons per square mile. From 1970 to 1976, mobile homes accounted for 70 percent of new housing units.

Not only are health problems associated with this crowding of the population into small areas of land, with a consequent overloading of sewage and drainage systems, but there is also increasing concern about health problems from "indoor pollution" in trailers. In many parts of the country, high levels of formaldehyde gas have been detected in mobile homes, emitted from the resins used in wood construction and from insulation. Health problems associated with formaldehyde range from respiratory ailments to cancer and birth defects. The latter are of particular concern, when so many young families start out in mobile homes, for lack of alternatives.

While mobile homes have financial advantages for families, for their community the reverse is true. Mobile homes generate lower property taxes for county revenues than do conventional homes, since they are taxed as personal property. Yet they demand at least as many services as do conventional homes.

To date, local, state and federal agencies on the whole have failed in their policies to recognize the contributing role which land ownership plays in the housing problem. Without adequate intervention on their part, housing policy in the region is largely shaped by the presence and powers of the corporate and absentee landholders who limit or define the alternatives to the status quo. There has been a growing regional frustration with this situation. In many areas of the Appalachian coalfields the income of miners has increased substantially during the past decade. Yet, even with larger incomes, many miners have been unable to obtain even small plots of land, making the building of one's own home an impossibility. Likewise, land is generally unavailable for builders and contractors; thus, there are few single homes or new subdivisions on the market. The experience of those trying to get federally funded low income housing units built in central Appalachia for the region's large number of elderly or low income families parallels the experience of the region's blue collar workers—quite simply, little land is available for housing. As land ownership patterns in the region continue to stifle both individual initiative and institutional efforts to solve housing problems, frustration mounts.

Census after census has revealed that the need for housing in Appalachia is a critical and long unaddressed problem. The region's chronic housing problems are likely to be greatly compounded in the coming years, particularly in the coalfields where more and more miners are needed to deliver the nation's energy resources. In West Virginia alone, according to the West Virginia Housing Development Fund, 85,000 new homes are needed before 1990 in the state's eleven southern coal counties—[21] where the concentration of landownership in a few hands is among the greatest found anywhere in this study. Here and elsewhere "boom towns" will exacerbate the present situation, as new mines are opened or as synthetic fuel plants are built. In Campbell County, Tennessee, for instance, where already over 50 percent of the housing is considered substandard, Koppers Company, which owns some 34 percent of the county, plans to build five synthetic-fuels plants. According to government studies, one plant alone can generate the need for 10,000 new workers.[22] It is anticipated that the housing problems in the noncoalfield areas of the region will also intensify if the current trend of migration into the region continues.

If the housing needs of Appalachia are to be met, new and creative solutions must be implemented by government agencies in partnership with citizens' groups who represent the landless majority. Strategies such as the use of eminent domain, just taxation for large corporation owners, land use planning with housing and quality of life issues as its cornerstone, innovative use of zoning, rebuilding on previous housing sites, protection of the interest of year round residents in counties with substantial second home development, etc., must be tried.

SEVEN

Ownership, Energy and the Land

Clearly, almost any use of the land will affect it. But, in Appalachia, no other use brings effects so pervasive and so permanent as those of energy development. The legacies of mining, especially strip mining, are well known. Other new developments in energy extraction—synthetic-fuel development, oil and gas, shale oil, pumped storage schemes—also will have impacts on the land itself. Now, more than ever, the costs the region is being asked to bear in order to meet national energy demands will be very long-term indeed. The short-term gains of strip mining for coal may preclude future extraction of deeper-lying coal. A stream may take several generations to renew itself after pollution by acid mine drainage. Renewal of mountaintops removed to extract their underlying coal will take billions of years—geologic rather than human time scales.

With new ownership patterns discussed in previous chapters, come new forms of technology that will have as far-reaching effects upon the land as those before it. These technologies cannot be considered in isolation. They too are influenced by ownership patterns. Clearly, an owner without the capital of Occidental Petroleum, through its subsidiary Island Creek Coal Company, would not undertake to plan a 68,000-acre mountaintop-removal strip mine. Nor, unless that land was held in a large block would it be likely or able to plan development on such a scale. The introduction of synthetic-fuel development by the big oil companies is also made more possible by their ownership of vast coal and land resources. While technologies of energy extraction are by no means governed by landownership patterns, the use of certain technologies at certain times and places is influenced by them.

NEW OWNERSHIP PATTERNS

As discussed in Chapter 2 on landownership, the structure of the coal industry in Appalachia changed dramatically during the 1960s. Some of the region's largest coal companies were acquired by oil companies—Pittsburg and Midway Coal by Gulf Oil in 1963, Consolidation Coal Company by Continental Oil in 1966, Island Creek Coal Company by Occidental Petroleum and Old Ben Coal Company by Standard Oil of Ohio in 1968. Other oil companies (for example, Exxon, Mobil, Texaco, and Ashland Oil) began to acquire smaller coal companies and coal reserves. In the 1970s, big oil and gas corporations extended and consolidated their control of Appalachian coal reserves.

With their increasing control of coal resources, the oil companies bring to the development of the region's coal a global decision-making context, and an unprecedented scale of capital and technical resources. Altogether in the survey counties, eleven oil and gas companies own approximately 1,239,698 acres of surface land and mineral rights. Two of the biggest oil companies—Continental Oil and Occidental Petroleum—own a total of 422,320 acres of surface and minerals combined in the survey area, and control thousands more acres through leasing.

Some of the local effects of this broad picture may be sketched in: In Logan County, West Virginia, more than 35,000 acres of coal reserves are now owned by oil companies, and a further 24,000 by Columbia Gas.[1] The Crystal Block coal mine and its accompanying coal reserves in Mingo County have been sold by U.S. Steel to Standard Oil of Ohio, together with two U.S. Steel mines in Pennsylvania. At $750 million, this was one of the largest business deals in coal history.[2] Allied Chemical Corporation's mineral holdings in Fayette and McDowell counties, West Virginia, have been absorbed into the larger holdings of Armco Steel and A. T. Massey (a subsidiary of St. Joe's Minerals of New York, now in association with Royal Dutch Shell).[3] Altogether in the fifteen-county sample in West Virginia, eight large oil companies were found to own more than 340,000 acres of minerals and over 50,000 acres of surface land.[4] In Tennessee, a family-held coal mining and landholding company, the Tennessee Consolidation Coal Company, has also been purchased by St. Joe's Minerals of New York, and incorporated in their recent agreement for joint development of coal resources with Royal Dutch Shell.[5] In Virginia and Kentucky, the properties of Virginia Iron Coal and Coke Company were purchased by Bates Manufacturing Company, and shortly afterward by American Natural Resources Corporation, a diversified energy corporation from Detroit, which is pioneering synthetic gas manufacture from coal in the Dakotas.[6] In eastern Kentucky, 60,000 acres of mineral rights previously owned by National Steel have been purchased by General Electric, a subsidiary of Utah International.[7] In 1979, a tentative agreement was signed for the Blue Diamond Coal Company of Knoxville, Tennessee, one of the largest of the remaining independent coal companies of the region, to be acquired by Amoco (Standard Oil of Indiana).[8] The deal was later dropped by Amoco, in part because of the uncertainties surrounding Blue Diamond's lease-holdings in Kentucky.

Our study also indicates that outright purchase of coal companies and lands does not tell the full story of the extent of oil-company control of coal resources in Appalachia. Leasing of mineral rights is extensive, and constitutes such a control of options for the use of land as to be de facto ownership. In West Virginia, Virginia, Kentucky, and Tennessee, our state reports conclude that leasing is a significant mode of control and development of coal resources. Leasing by absentee corporations is connected with absentee ownership. Review of courthouse transactions found a tendency of large, absentee corporate owners to lease their coal lands only to other larger absentee corporations. This is demonstrated most clearly in Martin County, Kentucky. There, the largest landowner in eastern Kentucky, Pocahontas-Kentucky Corporation (a subsidiary of Norfolk and Western Railroads), leases 10,116 acres of its coal reserves to Island Creek Coal Company (subsidiary of Occidental Petroleum); 5,256 acres to Wolf Creek Collieries and 12,408 acres to Martin County Coal Corporation (both subsidiaries of St. Joe's Minerals, and included in its agreement for joint coal development with Royal Dutch Shell); 17,870 acres to Webster County Coal Corporation and 13,400 acres to Pontiki Coal Company (both subsidiaries of MAPCO, Inc, of Tulsa, Oklahoma); and 16,164 acres to Ashland Oil Company through its subsidiary, Addington Brothers Mining. Nearly 95 percent of the coal owned by Pocahontas is leased to oil conglomerates.[9]

Cooperative ventures between large corporations are another means of extension and consolidation of their control of energy resources. The joint venture between St. Joe's Minerals and Royal Dutch Shell through its subsidiary Scallop Coal Company for joint exploitation of their coal resources is a case in point. In West Virginia, Exxon and Columbia Gas have pooled their property and resources in the new Monterey Mines in Lincoln County. The same two companies have also joined with Pennzoil in a secondary oil extraction project in the old Griffithsville Oil Field.

The increasing control of the region's coal resources by absentee energy conglomerates provides the capital and technical resources for ever-larger-scale technologies to be applied to the extraction of Appalachian coal. Strip mines extending across thousands of acres, removal of entire mountaintops, processing of coal into synthetic oil and gas—all have extensive impacts on the land and water, as well as on the lives of people in the region. At the same time, this form of ownership of the coal resources removes ever further from the possibility of local influence the decisions over the development of those resources (See discussion in Chapter 4). Care for the land is not the major concern of such corporations, which juggle international energy markets and resources to draw the greatest profits. As the Harlan County conservationist, with USDA's Soil Conservation Service, told us: "A private owner will use something, take care of it and keep it. But a large corporation doesn't have the same feelings. Nearly all of these corporations are absentee and their purposes are exploiting the land. When the coal is gone, there won't be much left."

The energy crisis is stimulating development of coal resources that lie on the fringes of the traditional coalfields of Appalachia. In central and northern West Virginia, southwest Virginia, southern Tennessee, and northern Alabama, our study found evidence of acquisition and consolidation of mineral resource control and the beginnings of coal development. In some counties (like Randolph in West Virginia and Walker and Tuscaloosa in Alabama), coal mining has been taking place in the past in conjunction with other forms of economic development (mainly agriculture). The impacts of past coal development have been mitigated by these counties' diversified local economies. The new scale of developments in these areas is likely to change their economic base (through restricting agricultural use of land, for example), and thus may intensify the impacts of energy development. In other counties (like Scott County, Virginia, Dekalb County, Alabama) coal mining has been barely existent and the impacts which accelerating leasing and buying of minerals will bring are new, though perhaps not welcome.

Our study found that the pattern of absentee ownership and control of mineral rights that has long characterized Central Appalachia is now extending into these fringe areas. The big oil companies are playing a significant role in the new wave of leasing and purchasing activity there.

In Braxton, Nicholas, and Webster counties, West Virginia, Sun Energy Corporation of Pennsylvania (tenth largest oil company in the United States) purchased 30,000 acres of mineral rights. Exxon made extensive purchases of minerals in central West Virginia counties through its subsidiary, Carter Oil. It also leased a reported 100,000 acres of mineral rights in Braxton, Nicholas, and Clay counties. In Randolph County, West Virginia, Amax, a diversified energy and minerals company, leased thousands of acres of mineral rights from the McMullen family. Other large energy corporations, like Mobil, Occidental Petroleum, and DLM (a subsidiary of General Energy Corporation of Lexington, Kentucky) hold extensive leases of coal in central West Virginia.[10] In Scott County, Virginia, a traditionally agricultural county with a pattern of mainly small landownership, rumored coal speculation was apparently taking place, although little hard information could be found in the county's deedbooks. Consolidation Coal Company (subsidiary of Continental Oil) was apparently leasing many acres of minerals and planning a new deep mine at Dungannon, but keeping its plans well out of the public eye.[11] Virginia Iron, Coal and Coke, now owned by American Natural Resources, owns over 1,500 acres of mineral rights in Scott County, and is involved in plans for a synthetic fuel plant at either Dungannon or Mendota.[12]

In Tennessee, the southern Cumberland Plateau is the main area of new coal speculation. While some coal mining has taken place in the past in this area, it appears that new-scale developments may soon affect it.

Plans by Amax to develop a 10,000 acre strip mine around Piney, in Sequatchie County, were shelved after much public protest in 1976, but residents are not convinced that they have been dropped. When the coal market picks up residents expect to see further attempts to strip mine their coal.[13] Gulf Resources and Chemical Corporation of Houston, Texas, leased more than 5,000 acres in the eastern part of Cumberland county, and adjoining acreage in Roane and Morgan counties, for large-scale development. Their 1977 Annual Report states "the location is favorable with respect to possible barge shipments to Europe and Japan." The Tennessee-Tombigbee Waterway, under development by the U.S. Army Corps of Engineers, would presumably be the route for such shipments, and it appears that the waterway will play a significant part in the development of southern Tennessee's coal reserves.[14] In northern Alabama, agricultural counties also along the Tennessee River like Dekalb and Marshall are also seeing coal speculation occurring. According to the Dekalb County probate judge, there was a significant amount of mineral buying and leasing, in the late 1970's and simultaneously, an increase in strip mining for coal in the county.[15] The Tennessee-Tombigbee Waterway may also have a role in the development of lignite resources further south in Alabama and Mississippi for possible lignite development. Phillips Petroleum has been the leading company in this leasing, with other big oil companies like Continental Oil also involved. Proposals have been made for synthetic-fuel plants in the area, using lignite as a feedstock.[16] With the expansion of absentee and corporate control of minerals into these new areas, it is likely that the "Appalachian Experience" of coal development will spread into formerly agricultural counties, leading to great changes in landownership and land-use patterns (see Chapter 6 on landownership and agriculture).

When looking at mineral rights speculation in Appalachia, one can no longer look only at coal. The Eastern Overthrust Belt, running northeast-ward from Alabama through Pennsylvania and into New York, is fast becoming one of the country's hottest prospects for oil and gas. The latest energy crisis combined with some big finds (Columbia Gas Systems brought in one of the biggest natural-gas test wells ever in Mineral County, West Virginia, in 1979; oil strikes have recently been made in Tennessee, Kentucky, and Virginia) to spark a new wave of oil- and gas-rights leasing across the region. While in the early stages of a "gold-rush" like this it is common to find a number of individual entrepreneurs and independent operators active, big oil and gas companies have extensive leasing of oil and gas rights in the area, and are actively expanding and consolidating their holdings. Standard Oil of Indiana (Amoco), for example, is reported by the *Wall Street Journal* to have 2.75 million acres of oil and gas rights in the Eastern Overthrust Belt, and has spent $25-30 million in leasing land and doing seismic tests.[17] Exxon Corporation has drilled several dry wells in Hardy County, West Virginia; Columbia Gas, which holds 348,777 acres of mineral rights in our survey counties of West Virginia, has drilled several wells in addition to its big strike of 1979. Gulf Oil Corporation and Atlantic Richfield Company (Arco) have agreed to a joint venture to explore 1.2 million acres in the Appalachian Basin. Arco will spend up to $26 million. Gulf is contributing most of the acreage. In Scott and Wise counties, Virginia, Penn Virginia Corporation, an independent drilling concern from Philadelphia, has said that it and other companies will drill 260 wells on more than 132,000 acres.[18] While in east Tennessee it is still possible for small independent operators to sink a well and hit it rich, the game is mainly and increasingly in the hands of the big companies that have the capital resources to do the seismic exploration, test wells, pipelines and the rest, and to withstand a succession of dry holes.[19]

The Eastern Overthrust Belt is only in part synonymous with the coalfields of Appalachia. In much of Virginia and West Virginia, drilling for oil and gas is taking place in areas outside the coalfields, which have been removed from the impacts of energy development in the past.

The search for resource independence is not confined to coal and oil, or even to energy resources generally. In Appalachia, new minerals are beginning to assume importance. Uranium is the one most obviously connected with the energy crisis, but other metals are beginning to be found and developed in the region. These may afford other industries independence from the increasingly complex political implications of resource extraction from Third World countries. OPEC is the most successful example of a Third World cartel to control Western access to scarce natural resources, but others have been attempted. Metals such as bauxite, chromium, and copper, which once were both cheap and readily available, are beginning to involve multinational corporations in political and economic costs they do not care to incur. In this world context, any "home" sources of such minerals may provide an independent supply that can be valuable to American corporations.

The mountains of western North Carolina and southwestern Virginia are important areas in the search for new minerals. Uranium exploration is currently taking place in national forest land around Grandfather Mountain in Avery County, North Carolina, and in the Devil's Fork area of the Jefferson National Forest in southwest Virginia. A survey by two University of North Carolina geology professors pointed to several areas of uranium deposits in the east, of which the most extensive run along the granite chain of the Appalachian Mountains.[20] They have predicted that within the next ten years, uranium mining will begin in one or more of these locations. So far, the country's experience with uranium mining in the West does not suggest that this new development for Appalachia will be entirely welcome. Strip mining is the most common method of extraction of the uranium-bearing deposits, and the devastating effects this method can have on the land and water in Appalachia's steep terrain are already known. Milling of the ore to extract the uranium from the rock involves crushing it to a fine powder then mixing it with sulfuric acid. Large volumes of wastes are entailed with this milling process, wastes that emit radioactivity for many years as one radionuclide decays into another.[21] Dusts from the piles of waste "tailings" in the West are carried for many miles on the winds, contaminating water, plant, and animal life. In Appalachia, the denser human population means more people will be exposed to contamination from such sources unless the operations are very strictly controlled. Rainwater may leach radioactive elements such as radium and thorium from the waste piles, contaminating surface- and groundwater supplies.

Another mineral whose exploitation is beginning in parts of Appalachia is bauxite. One company in particular, Gibbsite of New York, has been trying to mine bauxite here for ten years.[22] It bought up mineral leases for an estimated 15,000 acres in Ashe, Alleghany, Surry, and Wilkes counties in North Carolina and Grayson and Carroll counties in Virginia, before public outcry over its plans for surface mining of bauxite made it shift its test mining to another location. Recently the company announced new plans for bauxite mining and ore processing in Grayson County, Virginia, despite public protest. Bauxite is used in the manufacture of aluminum, and supplies on the international market are becoming increasingly uncertain with political instability in Central America and the Caribbean.

In Madison County, North Carolina, we have reports of new plans for extraction of bauxite, and also such minerals as barite, used in drilling oil wells; monazite, which is associated with the radionuclides cesium and thorium; and olivine, a chromium substitute which is used in making fire brick.

Other minerals besides coal have always been mined on a relatively small scale in parts of Appalachia. Zinc, manganese, feldspar and mica have all had local importance in various parts of the region. It appears that these are now being joined by a new wave of speculation in minerals which may become equally important in some local economies.

Changes in the ownership patterns of energy resources in Appalachia, which are summarized above, imply many new impacts on the land and water of the region. Increased coal production, and larger-scale mines, will intensify the effects of strip mining on the land and people that have already been experienced in the coalfields, and may extend these effects beyond the traditional coalfields. The conversion of coal into synthetic oil and gas will bring new environmental effects, few of which have been experienced in the region before. The extraction and processing of oil shale will also bring new impacts, mostly in areas outside the Central Appalachian coalfields. And the use of the region's abundant water supplies to supplement nuclear energy, through pumped-storage schemes, involves more destruction of farms and communities to meet energy demands.

The region has already witnessed conflicts between citizen and environmental groups and the coal companies. In the past ten to fifteen years, strip mining for coal has met with citizen resistance through every possible means. Our study suggests that in the future, such battles will have to be fought with new protagonists (big oil companies as well as independent coal companies, for example), over new environmental impacts (synthetic-fuel plants and oil-shale retorts, as well as strip mining on a larger scale than ever before), and in new areas (the fringes of the coalfields, the Knobs of Kentucky, the Blue Ridge of North Carolina, as well as in the older coalfields). The citizens' groups that seek to give local residents a voice in how their local resources are developed now face bigger battles. They face them in a national political context in which the need for energy often is given more weight than the social and environmental costs of energy development.

STRIP MINING

Perhaps no issue in Central Appalachia has been more emotion-laden than strip mining. While citizens have protested by every conceivable means, from lying down in front of bulldozers to lobbying for stricter governmental regulation, strip mining has only increased throughout the coalfields. As a national energy crisis demands independence from foreign oil, even greater amounts of coal are expected to be mined.

In West Virginia, the amount of strip-mined coal increased by almost 130 percent between 1960 and 1978, while deep-mine production fell by 42 percent. By the end of the period, strip-mined coal accounted for almost a quarter of all the coal produced in West Virginia. In eastern Kentucky, in 1960, only 13 percent of total coal production was from strip mining. By 1975, 53 percent of all coal mined in east Kentucky was strip mined. In seven of the survey counties in east Kentucky, over 70 percent of total coal production came from strip mining in 1977.[23] In Virginia, the same picture is presented: in 1978, a third of total state production of coal was strip-mined, over 10 million tons.

Some counties of the survey show an even more dramatic expansion of strip mining, which has had far-reaching effects on the land and people. In Wise County, Virginia, strip mining is the second largest land use in the county, after forestland. As of August 1979, over 10 percent of the total surface area in the county had already been stripped, more than 30,000 acres.

In Mingo County, West Virginia, strip-mine production increased from 104,570 tons in 1960 to 413,372 tons in 1979. Martin County, Kentucky, has also experienced a dramatic increase in stripping. By 1978, some 6,-126,461 tons of strip-mined coal were produced in Martin County, twice as much as was deep-mined.

As long as it remains economically attractive to do so, strip mining will continue at least on this scale in Central Appalachia. Indeed, current ownership and leasing patterns suggest that even more extensive tracts will be stripped. In areas like east Kentucky, some large landowners are attempting to consolidate their surface and mineral holdings in order to avoid surface owners' protests over stripping. In West Virginia, Island Creek Coal Company has announced a twenty-five-year plan to strip 68,000 acres on the Mingo-Logan County line, the largest strip mine in the East. It also has initiated an 8,000-acre strip project in Upshur County. On the Cumberland Plateau of southern Tennessee, Amax announced plans in 1976 to strip mine an initial tract of 10,000 acres. Further acreage was expected to be stripped later. The plans were dropped after challenges from local citizens through Save Our Cumberland Mountains led to water-quality permits being denied. However, residents suspect they have not heard the last of the plans.

Such large projects can only be contemplated because of concentrated land ownership patterns—if Island Creek or Amax had to get agreement from thousands of small landowners, they probably would never be able to start such a project. The transfer of mineral rights from small, independent coal companies to large, multinational energy companies also affects the scale of coal extraction in Appalachia. The president of Amherst Coal Company, largest of the locally owned coal companies in West Virginia, summarized his exasperation with big oil. Referring to Exxon's multimillion-dollar twin mine in Lincoln and Wayne counties, he said, "No commercial coal company would have dreamed of an expenditure like that." Big oil has undreamed-of capital available. Furthermore, the worldwide context in which it makes its decisions about development of the various energy resources it controls may make it independent of traditional considerations of labor supply, transportation costs, even market demands, which constrain independent coal companies.

Strip mining has a number of effects upon the land when it is conducted in steep terrain. Its disruption of the land in turn affects water supplies and quality, and, through such consequences as flooding, disrupts communities. While these impacts have been widely discussed and studied elsewhere, it is important to summarize some of them here.

By denuding vegetation and eroding top soil, strip mining reduces the capacity of the land to absorb rainwater, thus increasing peak flows in streams below strip-mined hillsides. Many studies have documented this effect of strip mining. The Beaver Creek Study, conducted by the United States Geological Survey from 1956 to 1966, monitored stream flows from two small watersheds in McCreary County, Kentucky, one of them mined, one undisturbed. Peak discharges from the mine watershed were consistently higher than from the unmined one (as much as one and a half times higher), and occurred more rapidly after rainfall.[24] The New River studies, conducted by the University of Tennessee gave rise to a computer model to predict the effect of strip mining upon flooding. The model predicts that a 5 percent disturbance of the watershed will produce a two- to four-foot increase in the 100-year flood state.[25] Both the Beaver Creek and New River studies show that even a small amount of land disturbance from strip mining (less than 10 percent) can greatly increase the amount of runoff and peak flow discharge during storms. A series of studies by the U.S. Forest Service Northeastern Experiment Station in east Kentucky comes to similar conclusions—"Peak flow rates increased by a factor of 3 to 5 after surface mining. Lag time was reduced, thus effecting an increase in the rate at which flood peaks move downstream. It appears that peak flow is directly and positively correlated with the percent of area disturbed during surface mining."[26] The one study that has been seized upon by the coal industry as apparently vindicating strip mining is subject to question. U.S. Forest Service engineer Willie Curtis issued a report in 1977 that compared 50 percent mined and undisturbed watersheds in Breathitt County, Kentucky, and Raleigh County, West Virginia.[27] He found that peak flows after the storm of 4–5 April 1977 had been higher in the undisturbed watersheds. Curtis suggested that a "sand-dune" effect may be operating, such that extremely disturbed land may hold large quantities of water in its broken-up rock. It has not been established that the sand-dune effect will occur in all cases of extreme devastation, or that it can be maintained over time as disturbed land settles and the spaces for water storage are reduced.[28] And the sand-dune effect probably does not operate in the more common situations where a smaller proportion of a watershed have been stripped.

Curtis's arguments raise another specter: if strip-mine spoil retains large amounts of water, it is also subject to the stress of that great weight.

Where slopes are steep, landslides could result, with even greater devastation of downstream areas. It is for this reason that strip-mine regulations seek to ensure that water does not seep into replaced overburden. But in their turn, these regulations imply increased runoff—a Catch-22 situation.

Strip mining erodes soil and hence contributes to increased sedimentation of streams. As creeks and rivers silt up, their carrying capacity is reduced and their likelihood of flooding is increased. Again, many studies document the connection between strip mining and increased sedimentation. The Evironmental Protection Agency estimated that for a certain degree of slope, active strip mines yield 2,000 times as much sediment as forest land of similar size and character.[29] The Stanford Research Institute report on surface coal mining in West Virginia found that in areas with generally steeper slopes and greater natural sedimentation, suspended sediment in strip-mines watersheds is more than 1,000 times that in similar drainage basins where there has been no significant mining.[30] Both the Beaver Creek and U.S. Forest Service studies in Breathitt County, Kentucky, similarly found a clear relationship between strip mining and sedimentation.[31] The U.S. Geological Survey and Army Corps of Engineers studies to determine the reasons for excessive sedimentation of Fishtrap Lake in east Kentucky characterized strip mining as the major contributor of unanticipated sediment.[32]

These scientific studies now confirm what Central Appalachian residents have known for many years. Strip mining causes significant damage to the land and in turn contributes to the frequency and severity of flooding. The Kentucky Department of Natural Resources report on the 1977 flood concludes, "Considering all the information on the effects of surface mining on runoff and erosion, small tributaries with a high percentage of recently disturbed land probably had a significantly higher flood level as a result of the surface mining.[33] Devastating effects of the flooding that has taken place in Appalachia in recent years following the strip-mine disturbance of the land were found in many communities we studied. In Mingo County, West Virginia, for example, the highest flood in the history of the Tug Fork River occurred in April 1977. According to Army Corps of Engineers report, total assessable damage done by the flood was approximately $200 million. More than 4,700 homes and 670 businesses were damaged. Six hundred homes were destroyed. Over 200 miles of highways and railroads were washed out.[34] By some miracle, no one was killed in the flood itself, although the shock, fear, and grief of the flood, and the strain of losing homes and belongings, took their toll after the flood, especially on older people. In addition to the direct physical losses, businesses in the area were closed for an extended period. Loss of sales and output was estimated at close to $11 million, and business losses resulting from the temporary closing of coal mines exceeded $30 million.

Flooding in the valley of the Tug Fork watershed has increased steadily both in frequency and height during the last thirty years, according to a report by the Tug Valley Recovery Center. Strip mining for coal in the valley has increased at a parallel rate and volume during that same period, while the average rainfall and the severity of storm events for the Tug Fork Basin area has remained constant.[35]

Elsewhere in Appalachia, areas that had never before had major floods began to be flooded after strip mining commenced in their watersheds. Camp Creek in Pike County, Kentucky, one such area, was devastated by floods in June 1979. Seven houses were washed downstream, one with two women inside. Heavy strip mining had begun on the head of the creek in 1975, and by 1979 the upper sections of the watershed had been completely strip-mined. While residents of Camp Creek had little hesitation in connecting this strip mining with their flood, government representatives denied any connection. "My dad's 85 years old, and if his father were alive he'd be 125, and they've lived in this hollow all their lives. There's never been anything like this in this hollow for 125 years. . . . The strip mines are just about two miles on up past us. . . . They don't care, just that lump of coal."

Not only has flooding become more frequent, higher, and more extensive since the advent of large-scale strip mining, but its effects are more destructive. The regular flooding of bottomland once enriched the soil by adding fertile silt. Now flooding deposits clay and acid materials from strip-mine operations, destroying agricultural land. As Becky Simpson, a resident of Cranks Creek, one of the most flooded areas of Harlan County, says, "Folks can't raise a garden and they can't farm anymore because clay mud has washed over the soil."

Coal mining's other impacts on the land and environment include its effects on water. Both deep and strip mining create acid drainage, which can destroy fish life in streams and make water unfit for drinking. Acid mine drainage is formed when toxic materials, generally pyritic minerals, are exposed to air and water. The pyrites are altered by oxidation to soluble sulfuric and iron compounds.[36] These salts dissolve in water to form sulfuric acid; and this in turn dissolves other minerals exposed by mining operations, such as nickel, aluminum, manganese. Some of these are toxic, others carcinogenic.

Appalachian coalfield streams are extensively degraded by mining practices. As energy development in the region expands, the problems may become even more severe. According to the 1978 Kentucky Water Quality Report to Congress, the entire eastern Kentucky region is plagued by low water quality, "indicative of the coal mining which takes place in the area." Pike County was found to be one of the worst affected—indeed, in a county twice the size of other east Kentucky counties, the Nature Reserves Commission was unable to find a single site suitable for a nature reserve.[37] A recent TVA survey shows the Powell River, running from southwest Virginia to the Norris Lake, to have "the most critical water quality problem in the [Tennessee] Valley, resulting from mining activities."[38] On the Cumberland Plateau in Tennessee, an area where strip mining for coal is likely to increase in coming years, a number of major streams have already been affected by acid mine drainage and sedimentation from strip mines. The "Plateau Muskie," an endangered fish species, has been all but destroyed in its once primary spawning grounds there. Wise County, one of the most heavily stripped counties in Central Appalachia, and other coalfield counties in southwest Virginia suffer the consequences in polluted streams and rivers. The Southwest Virginia 208 Water Quality Plan concluded that the seven-county coalfield area had nonpoint pollution problems caused by active and orphan surface mines.[39]

Strip mining can affect the availability of water supplies as well as their quality. Disruption of upper-level aquifers on the Cumberland Plateau has already affected the wells of residents near strip-mine operations, and may serve to lower the water table for years. Residents of Walker County and Dekalb County, Alabama, have also reported loss of domestic wells due to nearby strip mining. In such cases, drinking water may be completely denied local residents, as strip mining damages both surface water and groundwater.

Central Appalachian residents have now had enough experience with strip mining for coal to be well aware of its destructive effects. Lorraine Slone, a member of Concerned Citizens of Martin County (Kentucky), told us: "The earth was made to live on ... now, however, it is being destroyed in order to enrich the few at the expense of the many. The air and water are being filled with dust and chemicals, and the land is being ravaged by strip mining. Strip mining has driven off game and wildlife, has filled the streams with silt, and has increased water runoff on the hillsides, thereby increasing flooding. If this is kept up, there won't be a Martin County to worry about in twenty years."

The ill effects of strip mining on land and water have been widely acknowledged for some time, and gave rise to the 1976 federal strip-mine legislation to regulate strip-mine operations. However, the negative impacts of strip mining have not disappeared with the passage of this legislation. And, the legacies of past practices remain. "Orphan land"—unreclaimed strip mining—is widespread across Appalachia, and continues to wreak havoc with streams, fish life, and communities downstream. Public money is now being assigned to try to limit the damages caused by orphan lands, the sites of private profit. In Mingo County, West Virginia, about 7 percent of the county has been stripped, only about half of which has been revegetated. In Walker County, Alabama, much of the land mined before the federal act remains without seeding or grading. Unreclaimed land reputedly stretched "from one end of the county to the other." In Virginia, about 24,000 acres were stripped before the passage of the state's surface-mining law in 1966.

Nor is it certain that strip mining currently taking place under the aegis of the strip-mine law will have no deleterious effects on the land or water. Indeed, as the Tug Valley Recovery Center points out, "it is a virtual certainty that strip mining in steep slope areas will continue to result in hydrologic damage."[40] The federal regulations fail to address adequately some critical aspects of strip mining, including drainage control. And they do not consider the cumulative effects of stripping on a whole watershed. Furthermore, given the history of the industry's practices, it is unrealistic to expect companies to comply voluntarily with the new regulations. And the resources of the Office of Surface Mining to inspect sites on a continuing basis are quite inadequate.

Finally, there is a loophole in the federal law expressly designed to favor large-scale stripping operations. While strip mining along mountainsides is required to return the land to its original contour, removal of entire mountaintops is allowed. Only the large energy companies have the capital resources, equipment, and expertise to level an entire mountain—and they are increasingly the ones with the land.

SYNTHETIC-FUEL DEVELOPMENT

Plans for a national energy independence from imported oil include increased coal production not only for direct use of coal but also for conversion to synthetic liquid or gas fuel. While it appears that the main thrust of synthetic-fuel development will be in the West, where coal is cheaper, Appalachia will also have a role to play. Even a minor proportion of an $88 billion federal program will be a significant development for the region.

As federal dollars begin to become available under the new Synfuels Corporation, for feasibility studies, pilot and demonstration plants, and for financing commercial development, one may expect to see many more proposals for Appalachian sites. Already, plans have been announced and are underway to place synfuel plants in a number of the counties in our study: In Marshall County, Alabama, TVA has plans for a medium-BTU gasification plant to supply up to one-third of the energy needs of Tennessee Valley industry. The plant will produce the equivalent of 50,000 barrels of oil a day, using 20,000 tons of coal a day. Costs are expected to be in the $1-2 billion range, and construction is due to be completed in 1989. In Pike County, Kentucky, a low-BTU gasification plant is under construction, financed in part by the local government, in part by state, federal, and ARC funds. It will serve an industrial complex that has yet to be built, and a housing complex. The project has been beset by cost overruns and delays, as environmental controls have had to be added along the way. In Scott County, Virginia, Dynalectron, Inc. has preliminary plans for a liquefaction plant using the H-Coal process, to be sited in Dungannon or Mendota. The plant would process around 22,000 tons of coal a day, and would be a full-scale commercial version of the pilot now being run by Ashland Oil in Catlettsburg, Kentucky. Federal funds for a feasibility study have been applied for. In Wise County, Virginia, local officials have been lobbying hard for a synfuel plant to be located in the county, with a 628-acre site in St. Paul on the Clinch River earmarked for the project. As yet no definite plans have been secured. In Campbell County, Tennessee, Koppers Company, a major landowner identified in our study, has plans for a commercial-scale liquefaction plant to produce unleaded gasoline. In the final stage of development, up to six units would operate at the site, each producing the equivalent of 10,000 barrels per day. Some form of federal financing of the plant is expected, and a federal grant for a feasibility study of the Campbell County site and an Anderson County site has recently been awarded.

It may be expected that these proposals are only the beginning of a flood of synfuel development in Appalachia. As a Dynalectron, Inc. spokesman, William R. Dowling, has said, "The time is right for development of synthetic fuels, and we are proceeding hell-bent-for-leather on the projects."[41]

The impacts of large-scale synthetic-fuel development on the land and environment of Appalachia will not only come from the greatly increased strip mining of coal to supply the plants—although this will be a significant impact. Synthetic fuel plants themselves are expected to involve deleterious effects through toxic wastes and emission to air and water of toxic materials. They may constitute a serious health hazard to workers and to residents in neighboring communities.

Assessing the environmental impacts of a full-scale synfuels industry, and especially the consequences for human health, is speculative, for there are no commercial-scale or even demonstration plants that have been adequately studied to serve as a model. As the Department of Energy (DOE) points out: "First, the nature and quantities of toxic pollutants discharged to air and water or existing in the workplace or products must be estimated from fragmentary evidence; Second, the levels of pollutants must be related to the number and severity of health effects through highly speculative models and sparse data from experiments whose relevance is questionable."[42]

In the context of such lack of knowledge as to the safety of synfuels plants, one would expect the conservative approach to prevail, and slow and careful development to take place in order to avoid disastrous and unforeseeable impacts. However, the "energy crisis" and push for energy independence have prevailed over the voice of caution. As a result, some residents of the region fear that they will serve as "guinea pigs" for research on the environmental impacts of such plants.[43]

Enough evidence now exists to suggest that impacts on the environment and on human health are possible, indeed likely, from a synfuels industry. The plants will have impacts on land, water, and air, and through their emissions and final product, may affect the health of workers, neighboring residents, and consumers.

Synfuel plants will require large amounts of land for the plant site, for mining operations, and for disposal of immense quantities of solid waste. Indeed, their land requirements may constitute a restriction on siting in Appalachia, where the necessary flatland for a plant site is in short supply. The DOE siting study referred to above suggests that average plant-site needs for liquefaction plants range from 450 acres to 650 acres. The proposed Dungannon, Virginia, site is 470 acres, with a large additional area required for a buffer zone. The Campbell County site for the Koppers Company development is 1,600 acres; the TVA site in Marshall County, Alabama, 1,100 acres. In some circumstances where flatland is scarce, the large amounts needed for a synfuel plant would serve to deny the possibility of other industrial development in the community, including industrial development that would supply more jobs than the highly capital-intensive synfuel industry.

A further large area would be required to dispose of the solid wastes from a commercial-scale synfuel plant. The DOE has estimated solid waste generation from typical liquefaction technologies to be around seventy tons per hour—one railroad car full of waste every hour the plant operates (and they are expected to operate about 80 percent of the time). Disposal of such waste in a safe manner presents problems, since it consists mainly of ash and sludge that contain trace amounts of a wide variety of toxic and carcinogenic materials. Leaching of such materials would contaminate water supplies and render them unfit for drinking. Accordingly, landfills for the waste must be safeguarded from runoff, and leachate must be collected and disposed of separately. In many Appalachian communities, the danger of contamination from solid wastes of this kind would only add to already polluted water supplies from strip and deep mining.

All the synfuel processes consume large amounts of water. The hydrogen atoms of the water molecule are combined with the carbon of the coal to form the synthetic oil or gas. The DOE study estimates water needs of various liquefaction technologies to range from 6,000 to 9,000 gallons per minute (averaging 19 cubic feet per second). Gasification technologies require large amounts of water for cooling purposes, and consume three to five times as much water as the liquefaction processes. The ready availability of large amounts of water is considered to be one of the main attractions of the Appalachian region for synfuel development.

However, the large amounts required by synfuel plants may have significant impacts on supply at certain sites. For example, the Clinch River sites proposed for Scott and Wise counties, Virginia, may experience substantial losses of flow at certain times from the demands of a synfuel plant. The Clinch River in that area runs as low as 25 million gallons per day in times of drought (with an average low flow in summer of 40 million gallons per day).[44] Synfuel plants can consume 15 million gallons per day, or more, depending on the technology. Over half the flow of the Clinch River could thus be used up by a synfuel plant, severely reducing the availability of water to other users (including the expected population increase from the plant itself).

Synfuel plants may also have a significant impact on the quality of the region's water. Liquid wastes from a synfuel plant would be likely to include such pollutants as phenols, polycyclic aromatic hydrocarbons (PAH), trace metals and radionuclides. Possible health effects include cancers, liver damage, mutagenic effects and central-nervous-system damage.

The effects of synfuel plant discharges may exacerbate the problems already experienced in certain parts of the region from strip-mine and deep-mine pollution. The DOE study found several river systems in the study area to be problematic for synfuel siting because of existing water-quality concerns, including the Tug Fork along the Kentucky–West Virginia border, the Kanawha in West Virginia, the Licking, Kentucky, and Cumberland rivers in eastern Kentucky. More localized problems may also exist at other sites. However, the pressures to develop synthetic fuel are now so strong that they may override objections made on the basis of water quality.

Other impacts of the synfuel industry that are of concern to Appalachian residents include air pollution, and occupational health questions.[45] As with strip mining for coal, the benefits in jobs and profits favor a few, but the costs will affect many. Synthetic fuel processing is a capital-intensive industry, like the petrochemical industry, and relatively few jobs will be forthcoming for the money (including taxpayers' dollars), land, and other resources poured into these projects. Most commercial-sized plants, costing about $1-2 billion will require only a few hundred workers to run them. Construction crews numbering several thousand will descend upon the chosen community for four or five years, causing a temporary boom-town effect, then leave. Many of the permanent jobs will be highly skilled, and relatively few are likely to be open to local people. Local people will thus receive few of the benefits of this development, but will have to cope with the social costs (e.g., air, water, and land pollution). And when the plant has reached the end of its alloted life span (as little as twenty years), the local community will be left with serious residual problems and few resources to deal with them.

OIL SHALE

Included within the rubric of synthetic fuels, though not deriving from coal, oil from shale is considered one of the most promising new technologies to meet the energy demands. Until recently, interest was almost entirely in the West, but new exploration of oil-shale deposits in the East, together with new technical developments for extracting oil from eastern Devonian deposits, have given oil shale a significant potential for development in the East.

The Institute of Gas Technology estimates that the Eastern United States has some 420 billion barrels of oil in easily accessible shale formations.[46] One hundred ninety billion barrels of this are estimated to be in Kentucky, in a 2,650 square mile crescent east, south, and west of the Bluegrass region. Oil shale is also located in West Virginia and Tennessee. The DOE has initiated an Eastern Gas Shale Project, based in Morgantown, West Virginia, which is surveying for shale deposits in West Virginia, Ohio, Pennsylvania, and Kentucky. In addition, the DOE's regional office in Atlanta has applied for funds for a "full-blown resource assessment" of Tennessee's shale deposits.[47]

Such government interest is matched by private commercial interest. Woodstock Minerals, Inc., of Los Angeles has been seeking lands for shale development in Alabama and Tennessee. In Kentucky, the publicity surrounding the Addington brothers' leasing of oil shale has dramatically increased public awareness of the issue. After selling their eastern Kentucky coal business to Ashland Oil Company for a reported $113 million, one Addington brother, Robert, started leasing for oil shale in the Knobs area of northeastern Kentucky. The other brother, Larry, began leasing in central and south central Kentucky under the name of Addington Oil Company. His company managed to lease about 150,000 acres in counties to the south and west of the Kentucky River.

However, numerous residents in that predominantly agricultural area began to protest that the leases were fraudulent. They had been told that the leases would not permit strip mining, when in fact they allowed mining by any conventional method. They also maintained that they had been told that the leases were like the two-year oil leases with which they were familiar and which required renewal. In fact the leases were perpetual. After continued protests and the threat of a suit by the state attorney general, Addington Oil Company agreed to renegotiate or cancel leases. By midsummer 1980, over seventy percent of the lessors had canceled or renegotiated their leases. No more than a fourth of the lessors had retained the original leases.[48] Larry Addington has since assigned his leases to an Ohio company previously involved only in the stripping of coal.

Robert Addington's company, American Syn-Crude, has approximately 90,000 acres under lease in northeastern Kentucky and owns another 3,000 acres fee simple. The magnitude of this leasing is perhaps better illustrated by the acreage leased in particular northeastern Kentucky counties. American Syn-Crude has 40,000 acres under lease in Lewis County, 18,000 acres in Fleming, and 14,000 acres in Powell. Additional acreage is under lease in such counties as Rowan, Bath, and Estill. In the wake of the publicity involving Addington Oil Company's leases, twenty-nine property owners in Estill County successfully sued Robert Addington's company to cancel their leases.

The Addington brothers have not been the only actors on the Kentucky oil shale stage. Phillips Petroleum started leasing in Kentucky in 1981 and now has about 23,500 acres under lease in eighteen different counties. Breckinridge Minerals, a subsidiary of Southern Pacific Petroleum of Sydney, Australia, has leased more than 22,000 acres of oil shale land in Montgomery and surrounding counties. Sixteen thousand of those acres are in Montgomery County, representing almost 14 percent of the county's land area. The total land under lease for possible oil shale development in Kentucky probably totals less than 200,000 acres, though the state claims that at least 300,000 acres are under lease. This is out of a state total of some 1.4 to 1.7 million acres suitable for the strip mining of oil shale.[49]

The posture of the state government in Kentucky has generally been prodevelopment with respect to oil shale. The 1980 state legislature did, however, place a temporary moratorium on large scale oil shale development projects until environmental protection standards could be drafted.[50] The state evidently supports oil shale development as an economic development strategy and thus rationalizes the utilization of public funds for its promotion. Companies are informed that they can gain entry to the oil shale scene either through buying leases from already established companies or going directly to the local land owners. The prodevelopment attitude of the state government no doubt contributes to the continued interest in oil shale development in the state.

Shale-oil extraction involves significant environmental impacts on land, air and water. Two main technologies are being developed for its extraction: surface processing, which involves mining of the shale rock, processing at high temperatures in a retort, and disposal of the large quantities of solid waste generated; and "in situ" or underground extraction, which involves heating the shale while still in place underground, and piping up the extracted oil to the surface. Above-ground techniques for shale-oil extraction have been developed for some time, but have not been commercially viable or tested until now. Underground techniques have been developed mainly by Occidental Petroleum in the West, and are still some way from commercial stage.[51]

Surface extraction of oil from shale requires strip mining of the rock, with all the known environmental problems to land and water created by this method of mining coal. The rock would then be taken to a retort, creating potential dust and air-pollution problems. It would be heated to 900 degrees to release the kerogen, which then would most likely need to be refined further in a conventional oil refinery to provide fuel. Much water would be consumed in the process—some three to seven barrels of water for each barrel of oil produced. In the West, restrictions on the availability of water may place a ceiling on oil-shale development, and even in the East, demands for water by an oil-shale industry would be significant.

Water-pollution problems may be serious. Spent shale contains salts, including potentially toxic metals like boron, fluoride, and molybdenum, which could leach from waste-storage areas and contaminate surface water and groundwater supplies. Underground retorting of shale may avoid some of the other environmental effects, but could potentially damage groundwater supplies.

Impacts on the land from surface processing of shale to extract the oil include a significant waste-disposal problem. About one ton of rock yields a barrel of oil, and the heating creates a "popcorn effect" so that the spent shale has greater volume than the mined rock. In the West, it has been seriously suggested that a few unused canyons could be filled up and leveled with spent shale. In the East, disposal of the waste may be even more difficult, since the land is more densely populated. Wherever the site is, methods must be found to seal it so that leaching from the shale cannot take place.

Most of the environmental questions surrounding shale oil cannot be answered at the current level of technical knowledge. Environmental controls that work in the laboratory or in pilot plants may not meet the needs of commercial-sized facilities. To push ahead too fast with commercial development of untried and untested methods could have disastrous effects on the land and water of Appalachia.

As oil and gas prices rise, it becomes economically feasible to seek oil and gas in areas that had aroused little exploratory interest when prices were low and extraction costs high. Following the 1973 "oil crisis" there was a flurry of speculative oil and gas drilling in Appalachia, and the late 1970s saw a bustle of activity. In the so-called "Appalachian basin," there is potential both for shallow-drilled oil and gas wells and for very deep wells, a mile or more beneath the surface.

Currently there are producing oil and gas wells in parts of southwest Virginia (Lee County), east Kentucky (Letcher County produced over

220,000 barrels of oil in 1978), and east Tennessee (a total of 311 wells producing oil in Morgan, Scott, and Fentress counties). But the current picture of oil production is but a miniature of future prospects in the region. Exploration and leasing for oil and gas has extended from those counties which have long been known as potential producers into largely agricultural counties where oil and gas leasing is a novelty. In Cocke County, Tennessee, for example, on the North Carolina line, there were only twelve oil and gas leases recorded in 1979; in 1980 there were about 600. As much as 10 million acres may already have been leased in Appalachia, according to the *Wall Street Journal*,[52] and major oil companies like Exxon, Gulf Oil, and Standard of Indiana (Amoco) have an appreciable interest.

One reason for the increased interest in Appalachian oil is that returns on drilling investment, although modest, are more assured than in other areas. While in Texas, only 66 percent of wells drilled come up with commercially viable amounts of hydrocarbons, and in Kansas the proportion is 54 percent, 90 percent of wells drilled in Ohio, Pennsylvania, and West Virginia produce. A typical well will pay back its cost in three to four years, and return three times its initial investment in fifteen years.

In Tennessee, however, the picture is very different. Less than half the wells drilled come in, but a well can pay back its cost in as little as a week. There is also more unexplored acreage in Tennessee than in other Appalachian states, which is now attracting many "wildcatters" (operators who drill wells more than a mile from existing producing wells), independent operators, and investors looking for a gamble. Six hundred wells were drilled in Tennessee in 1979, a record for the state. More would be drilled if more gas pipelines were constructed to transport the gas that is often found in concert with oil.

There are now estimated to be some 5 million acres of oil and gas rights under lease in Tennessee. Phillips Petroleum alone has leased 123,000 acres in east Tennessee. Other big oil companies also have substantial leases. While Scott, Morgan, and Fentress counties are the main boom areas for exploration, leasing is also taking place further south, in Cumberland County and in counties east of Knoxville—Jefferson and Cocke counties in particular.

In Virginia there has been a similar increase in leasing of oil and gas rights in recent years, although little new drilling is taking place as yet. According to the Virginia Department of Labor and Industry, total acreage under lease at the end of 1979 was over 3 million acres, an increase of 68 percent from the previous year.[53] Six major oil companies—Amoco, Columbia Gas, Gulf Oil, Philadelphia Oil, Exxon, and Chevron—lease 79 percent of these acres. The potential oil and gas area extends from Lee County in the far southwest corner of the state, northeastward as far as Frederick County in the upper end of the Shenandoah Valley.

While West Virginia has a long history of oil and gas production, recent years have seen a surge in exploration and leasing of oil and gas rights. Again, the oil and gas companies have been active in the area: Columbia Gas is extending its leasing; Consolidated Natural Gas has extensive leasing in the eastern panhandle; Exxon has drilled a number of dry wells in recent years, and has a "significant lease position" in West Virginia through its subsidiary, Carter Oil. Amoco also has some dry holes and is doing seismic research in the area. Much of the new leasing and exploration is taking place in the north-central and northeastern parts of the state. In the Alleghany Highlands area, it is taking place in primarily agricultural counties that have not previously known the effects of energy development.

Western North Carolina is another area that in the past has been outside the energy development zones of Appalachia, but through oil and gas exploration and leasing is now being drawn into energy development. The concealed part of the Eastern Overthrust Belt, which tuns through Georgia, Western North Carolina, and up into Virginia, may have potential for yielding oil and gas through deep drilling (maybe a mile or more below the surface). The U.S. Forest Service has recently reported significant oil and gas leasing under national forest land in western North Carolina.[54]

Amoco has leased 122,000 acres in Cherokee, Clay, Graham, and Transylvania counties, and Weaver Gas and Oil Corporation of Houston has leased 120,000 acres in Cherokee, Graham, Madison, and Swain counties. So far the interest has been aroused from shock-wave soundings: exploratory drilling is not expected to take place for several years.

Oil and gas extraction is not normally regarded in Appalachia as being environmentally damaging, for few people have experienced it at first hand. However, as exploratory and commercial drilling is beginning to spread, residents are starting to encounter some of the possible ill effects on their land. One Randolph County, West Virginia, farmer found his pasture damaged with core holes, his road and fence destroyed. After one of his cows died from drinking water contaminated by runoff from drilling sites, he was forced to sell the rest of his livestock. A similar experience has been reported from neighboring Barbour County. One property owner was given only a day's notice that drilling for gas was about to start; his fence and fruit tree were torn down and a road bulldozed through his woods, destroying valuable timber. His farm pond was used as a water source for drilling operations, killing its fish. The county road leading to his farm was severely damaged.

Residents of Lincoln County, West Virginia, have also reported some of the ill effects of oil and gas drilling. A consortium of Pennzoil, Exxon, Columbia Gas, and Guyan Oil has initiated a project in the county, using water flooding and carbon dioxide gas under pressure for secondary extraction of oil from old wells. Preliminary work on ninety acres, before the consortium was sued for operating under federal funds without submitting an environmental-impact statement, resulted in ruined well water, polluted streams, torn up roads, and destroyed farmland.

Elsewhere in West Virginia, and in other areas where gas has been found, fires from gas wells have created a nuisance, air pollution and a potential danger to nearby homes. In the populated East, in contrast to the West, oil and gas wells have to coexist with communities, farms, and forests, and many more safeguards may be needed to ensure minimum damage to the land and environment.

PUMPED STORAGE FACILITIES

Coal and water have been traditional keys to Appalachia's energy development. The TVA was founded in the Depression on the basis of power generation through hydroelectric schemes (to be meshed with flood control and recreation provision), and only subsequently extended into coal-fired and nuclear power generation. Dams for electricity generation have been combined with dams for flood control to harness just about every river system in Appalachia. And recently a new use of water for energy production has been proposed, and met with stiff citizen opposition.

Controversy over pumped-storage facilities has been most pronounced in southwest Virginia, although an earlier proposal caused conflict in West Virginia. Appalachian Power Company, a subsidiary of the American Electric Company, has proposed a series of pumped-storage facilities in Virginia.[55] These would serve as giant storage "batteries" for electricity. At night, when power demand is low, surplus electricity would be used to pump water uphill from a lower lake to a higher one. In the day, when power demand increases, the water would be run back downhill through turbines to generate electricity. Any such scheme is inefficient, requiring about 4 kilowatts of electricity to pump uphill enough water to generate 3 kilowatts on its downhill run. And pumped-storage schemes would mainly be useful in conjunction with nuclear power plants, which cannot be turned down at night as demand lessens, rather than with coal-fired plants, which are quite flexible.

APCO began its long search for a pumped-storage site on the New River in Virginia. In preparation for its Blue Ridge Impoundment Project, APCO acquired some 12,000 acres in Grayson County, much of it prime agricultural land. After years of battles on a national and local front, Congress designated that section of the New River a "wild and scenic river," and the project was stopped. APCO is now realizing substantial profits from the resale of its Grayson County acquisitions. Undaunted by its defeat over the Blue Ridge project, APCO then announced two proposals for pumped storage schemes, on Powell Mountain in Scott County, and at Brumley Gap in Washington County, Virginia. They proposed the largest pumped-storage facilities in the Western Hemisphere, each capable of producing 3 million kilowatts of peaking power. Both plans have potentially significant impacts on the land and people.

APCO's Brumley Gap proposal involved the flooding of about one hundred homes, plus churches and stores, in order to make the lower lake. "Hidden Valley," up the mountain, would hold the upper lake, flooding a state game refuge, significant Native American archaeological sites, and obliterating one of the few streams where native trout remain. During the night the lower lake would be pumped dry to fill the upper lake. During the day water would drain from the upper lake to run turbines to generate additional power. Thus neither lake could support fish and wildlife.

APCO's Powell Mountain site is wilder and more remote. Most of it is within the Jefferson National Forest. Parts are being studied for designation as wilderness areas, to preserve some unspoiled natural beauty in an area where increasing strip mining for coal has scarred many hillsides. About twenty-five families would have been flooded out by the lower lake; a hundred more would have faced the uncertainties of living below the 300-foot earthern dam. The upper lake would take in the Big Cherry Reservoir, source of water for the town of Big Stone Gap. amd render the water unfit for drinking.

Beyond the impacts of the flooding and the hydroelectric machinery on the land, there are potential ill effects from the 765-kilovolt powerlines that would transport electricity to and from the sites. Ultra-low-frequency electromagnetic waves emitted by these lines are now strongly suspected of causing such health effects as stress, increased susceptibility to disease, even cancer.[56] In some areas of the United States, farmers have already experienced problems with grazing animals and growing crops under high-voltage powerlines—crops do not mature as usual, cows have difficulty letting down their milk. Honeybees have responded to electromagnetic waves by gaining less weight, producing fewer young, and losing their ability to withstand winter cold. Mice in tests respond to low frequency radiation by signs of stress, changes in blood chemistry, and increased infant mortality over several generations.

APCO's plans for the Powell Mountain site have been dropped, after vociferous local and environmental groups' opposition. However, Brumley Gap may yet see its farmlands flooded, its community destroyed, and families relocated, to make way for a pumped-storage facility. The formation of a coalition of concerned citizen groups—the Coalition of Appalachian Electric Consumers—resulted from opposition to APCO's plans, and the coalition continues to play a significant role in challenging the plans and policies of American Power Company and its subsidiaries.

These plans, and a plan by Alleghany Power System to build a pumped-storage facility in the upper Canaan Valley of Tucker County, West Virginia (the focus of environmentalists' protests,[57] raise some significant questions. The power generated would in each case be transported out of the region, to serve the peak needs of urban areas miles away. Here, as elsewhere in the region, citizens' groups are asking, What price must rural communities expect to pay in order to meet national energy demands?

During the last two decades in the Appalachian region, conflicts over the use and misuse of the land for energy development have been intense. Unchecked, unreclaimed strip mining, in particular, has provoked bitter grassroots outcry. For many local citizens, the concerns are not simply aesthetic ones. For them, strip mining destroys water supplies, endangers homes, takes away deep mining jobs, and erodes communities and a way of life. In response, to these and related grievances, a score of grassroots organizations have sprung up to voice their interests. Their efforts have been frustrated, among other things, by a deep-rooted attitude, locally and nationally, that landowners have the right to do whatever they please without public accountability, regardless of social and environmental consequences nearby. With the passage of state and national legislation on surface mining and other environmental concerns, the battle over whether regulation will occur has given way to battle about the extent of governmental regulations.

Within the traditional coalfields, as has been seen, strip mining and other energy developments are increasingly dominated by larger corporate units, primarily multinational oil and energy firms. With the consolidation of their control, energy investments will be on a bigger scale, with far-reaching impacts. Strip mining will involve thousands of acres at a time, rather than hundreds and will affect more people and communities. At the same time, decisions about where, when and how mining is to occur will be made further away from the reach of local citizens and officials, who will have to form coalitions with other similarly affected to let their voices be heard effectively.

While conflicts over energy developments may escalate in scale in the traditional coalfields, they are also likely to extend to new areas. This expansion of energy developments into new areas of Appalachia already has provoked response of citizens and officials who are not yet as economically dependent upon the energy industry as in the older coalfield areas. In these new areas, which often have relatively dispersed land holdings, farmers, local businessmen and others have been mobilized with more numbers and with greater effectiveness than in the sectors where landownership and economic development have been dominated by large, corporate energy owners. For example, the search for oil shale in Central Kentucky, the expansion of strip mining into Lincoln County, West Virginia, and Sequatchie and Van Buren, Tennessee, and the threat of pump-storage facilities in Washington County, Virginia—all agricultural areas not previously dominated by energy producers—have been met with well-organized citizens response.

Both in the "traditional" energy fields and in the "new areas," local communities will face environmental impacts growing from new technologies, such as synthetic fuels, and from the search for new minerals, such as uranium. While some of these impacts have been outlined in this chapter, complete information on the consequences of these new energy technologies is lacking. Some interests are pressing for full scale, rapid development of these energy sources. Local officials and citizens, however, more than ever before need to have a voice in this process to avoid the costly environmental and social consequences experienced with past energy "booms" in Appalachia. For their voice to be heard, government agencies, too, must recognize the right and the importance of local citizens' participation on matters related to the development and use of the land in their communities.

EIGHT

A Call to Action

For decades people in communities throughout the Appalachian region have been struggling against the concentrated, usually absentee control of the region's land and mineral resources. The Appalachian Landownership Study must be seen as part of that decades old struggle. Many of the citizens and scholars who became members of the Task Force or otherwise participated in the study were veterans of those battles. Too often, they had been hampered in their efforts by insufficient information about the control of land and minerals in the region. For them, this study was the chance to document in a comprehensive way landownership patterns in the Appalachian region and to analyze their impact on rural communities.

The importance of the study, however, does not lie in the documentation alone, as comprehensive and important as it may be. It comes also from the process by which the information was gathered, for whom it was gathered, and for what purpose it was gathered. Inasmuch as possible, local citizens were to do the research, they were to document ownership patterns primarily for themselves and their fellow citizens, and the information was to be used to further struggles for social change in the mountains. So, the study must be seen not only as the documentation of landownership patterns and their impacts, but also as a call to action that would alter both the patterns and the impacts. Toward this latter end, the Task Force formulated a number of recommendations and strategies for change that could serve to ameliorate the most adverse impacts in the short term and dramatically change the ownership patterns themselves in the long term.

One of the discoveries of the Task Force was that knowledge of landownership in the Appalachian region is much like the patterns of landownership themselves. Control of the information is highly concentrated in the hands of a few government agencies, land speculators, and corporations— absentee interests that are affected financially but not otherwise by what they find and how they use it. That control of information is apparent in every courthouse and state agency in the region. Much of the information that should be there for public inspection is either not there or not open to the public. Data on corporate mineral ownership and leasing often depends on the willingness of the corporations to report it. Some state regulatory agencies claim confidentiality for their records on utility ownership, a clear abrogation of the public's right to know.

It is difficult to develop rational policy options relative to landownership issues in the absence of accurate, complete, and public data on ownership. Although this study was able to document landownership patterns in eighty counties, there were continual problems with the accuracy, completeness, and availability of data in all the states in which the survey was undertaken. The problems result from two interrelated factors: the manner in which landownership, taxation, and use information is recorded and the type of reporting that is required.

Traditionally, collection of ownership and taxation data has been left to the counties, with technical assistance and implementing legislation provided by the states. The result is often inaccurate, incomplete, and confusing records that in effect conceal ownership and taxation information from the public. To deal with these unclear and/or partial ownership records at the local level, the recording of landownership and taxation information should be standardized, at least within states. Such action would help prevent the concealment of property ownership and taxation that maintains current inequities. Once such standard procedures are mandated, there should be such monitoring and penalties as are necessary to insure compliance by local officials as well as corporate and other land owners.

A further step may be necessary if we are to make available complete information on the ownership of land in Appalachia by energy conglomerates and other corporate entities. These national and multinational companies do not recognize state boundaries, their ownership usually transcending states and regions of the country. The establishment of a landownership census system or an inventory that would document landownership on a national basis would accurately document not only who owns Appalachia, but also who owns America.[1] Such a system would also serve the functions of affirming landownership and use as national issues as well as standardizing ownership information. Both are long overdue.

Another recurrent problem as Task Force members undertook their research was the widespread under-reporting of mineral properties. In Kentucky, for instance, it was hard to determine who, if anyone, knew the actual extent of corporate ownership of the mineral wealth of the state. Efforts by the State Department of Revenue had clearly demonstrated two things: that companies did not take the state effort seriously (many simply did not return the questionnaire) and that many who did report significantly under-reported their actual ownership. Even in states where reporting does occur, it is often impossible to tell precisely where the mineral rights are located. The establishment of programs requiring detailed reporting, recording, and mapping of all mineral properties is the minimum action needed. Penalties for failure to comply should be immediate and severe. Suitable penalties might be the reversion of those mineral rights to the surface owner or forfeiture of such properties to the state for future dispersal consistent with the development needs of the local community.

The adequacy of any of the above measures will depend on the extent to which information is available to public inspection. Current landownership and taxation records, as inadequate as they may be, are often not readily accessible to the public. Many states, including Tennessee and Kentucky, have now compiled landownership and taxation information at the state level, in their capacity of providing technical assistance to county officials. This information, however, has not been considered public information at the state level, forcing researchers and local citizens to search through county courthouse records. All ownership and taxation records compiled by any state agency should be a matter of public record.

Once adequate and readily accessible ownership information is assured, public policy options should become much clearer. Any policy actions, though, must be based upon a broad public awareness of landownership issues as well as citizen participation at all levels of decision-making. Upon completion of the study, the Task Force felt that there were three general policy options: impact mitigation, land retention, and land reform. The first two of these are premised on the expectation that, given today's political and economic climate, fundamental reform of landownership and use policies is unlikely in the near term. As a result, short term goals must deal either with the effects or symptoms of the basic problem or with the effort to prevent further concentration of ownership.

The adverse impacts of the dominant ownership patterns in Appalachia are well known and have been thoroughly documented in the study. The mitigation of these impacts will be a central component of citizen action in the region for the foreseeable future. It should also be a primary motivation behind public policy decisions affecting the region. Such actions do not address directly the underlying structure of landownership, but rather deal with the effects of that ownership on the region's citizens. However, they should be seen as steps in a lengthy process by which more fundamental reforms will be achieved. Policies must be implemented to: provide adequate tax revenue for the provision of services; promote diverse economic development; provide housing adequate to meet present and potential community needs; insure energy development that is nondestructive of local communities; and in general insure land use beneficial to the community as a whole.

The study found consistent patterns of underassessment of property, especially minerals; inequitable distribution of the tax burden, such that small, local property owners pay more than the large, absentee owners; and low payments for government properties in lieu of taxation. As a result of these patterns, county governments lack revenue to provide the basic services that their citizens have a right to expect. The counties must either do without needed services or turn to federal and state governments for additional revenue.

This pattern of underassessment was most obvious in Alabama and Kentucky, though it was also quite evident in Virginia, Tennessee, and elsewhere. The value of surface land recorded in the tax books was uniformly low, even though it was supposed to be appraised at fair market value. While one would assume that "fair market value" means the price a willing buyer would pay to a willing seller, there seemed to be no standard method for determining that value. In numerous cases, the appraisal could not have possibly reflected the actual value of the land. This was particularly true in the case of large corporate holdings. In cases where it has not already been done, a standard method for determining true and actual value should be established and uniformly applied.

The net effect of the failure to do so is to shift the tax burden to the smaller owners, most likely local residents who live or operate businesses on their land. Counties dominated by large scale corporate and/or absentee owners are penalized in two ways: first, because the large land holdings are usually underassessed, and, second, because the lands are usually held for speculative purposes and not developed in ways that would contribute to the local tax base.

There are, however, policy options by which to alter this situation. One option would include a progressive property tax system, such that the more land an owner has, the greater the assessment rate applied to it. This could be incorporated as a variation of the rate structures currently based on land use. Another option would be to place a tax on "excess acreage," i.e., land above a certain acreage or land held for speculation would be assessed at a higher rate. For example, a bill recently introduced in the West Virginia state legislature would add $2 an acre in taxes for holdings between 2,000 and 5,000 acres; $3 for 5,000 to 8,000 acres; $3.50 for 8,000 to 12,000 acres; and $4 for over 12,000 acres.[2]

The most dramatic failure of the property tax system in Appalachia lies in its reluctance to tax mineral reserves at anything more than a small fraction of their real market value. The reasons for this state of affairs are complex and include such factors as: inadequate knowledge about mineral ownership; inadequate knowledge about the extent of mineral reserves; difficulties in determining the fair market value of those reserves; and actions by large mineral holders to prevent fair and equitable taxation. All of these are relatively amenable to technical solutions with the exception of the latter. What seems to be missing is the political will on the part of state legislatures to enact a fair and equitable minerals taxation system and commit the necessary resources to implement it.

Once the ownership is known and locations of such ownership are mapped, the task remaining is to develop a system for determining fair market value. Minerals not held for exploitation could be exempted from taxes subsequently assessed on that true market value. Coal industry representatives maintain that attaching a fair market value to their unmined minerals is not possible, though they seem to have little difficulty doing so when they purchase or sell ther mineral rights. Experiences in West Virginia and other states indicate, however, that systems of taxing unmined minerals are indeed feasible and can yield significant additional revenue.

West Virginia is currently implementing such a system for taxing mineral reserves. Although their effort has been criticized as being too conservative in that it does not really establish the full market value of coal as a basis for assessment, it has added substantial revenue to the state's coffers. If the fair market criterion were applied, the revenue potential would be dramatic. An appraisal expert recently testified that West Virginia was losing more than $50 million dollars annually because of its failure to appraise coal at its real market value.[3] The Kentucky Fair Tax Coalition has estimated that if an unmined minerals tax were enacted in that state, the additional annual revenues generated would be $64.6 million with $43 million of that going to local school districts and county governments.[4]

The development of a uniform minerals taxation system based on fair market value is even more important, given the new mineral leasing and exploration now taking place in traditionally noncoal areas of Appalachia. Such counties are ill equipped to make the determinations necessary for taxing these new minerals. Thus, many of these mineral resources, particularly oil and gas, are not now being taxed adequately, if taxed at all. If the location and ownership of these resources were identified and taxed on the basis of fair market value, counties could realize dramatic increases in their tax revenues. For example, one estimate suggests that the state of West Virginia is losing tax revenue on oil and gas alone on the order of $30 million annually.[5]

Extensive federal holdings within a county also pose special problems for the county's ability to generate property tax revenue. The removal of land and minerals from the local tax base both diminishes the potential local tax revenue and places a heavy burden on other landholders in the county. In the case of state owned land there are no reimbursements for the tax loss to the locality; in the case of federal ownership (especially National Forest Service) there are in-lieu payments, which are now set at a minimum of 75¢ an acre. In most cases, however, the in-lieu payments do not adequately compensate the county for its loss of revenue. The Task Force suggests that in-lieu payments for government lands be maintained at a level that would approximate the lost tax revenue. For example, they might be increased to equal the average tax per acre paid by local owners of comparable property. Also, future federal acquisitions should not take place without the establishment of adequate compensation formulas that take into account the fiscal impacts of such property in the county.

Economic underdevelopment is a long-recognized problem in many sections of Appalachia. Though it has usually been explained in terms of such factors as isolation and the qualities of the indigenous labor force, this study has found that patterns of landownership are also important elements in the persistence of economic underdevelopment. The major impacts of landownership patterns on economic development stem from the lack of available land for industrial siting, lack of adequate infrastructure for such development, and the lack of sufficient local capital to fund such development. As a result, industrial diversification is virtually impossible and the labor force is at the mercy of the boom and bust cycles of the dominant industry.

Three strategies are available for dealing with the lack of land for either industrial siting or construction of the necessary infrastructure. In areas

Call to Action

where absentee corporate ownership limits the availability of land, tax incentives might be employed to encourage such owners to make land available for industrial development. If that were unsuccessful, a second strategy could be employed: state and/or localities could be empowered with the right to condemn the needed land in the interest of the economic development of the total community. In areas where federal ownership is dominant, a third strategy might be needed. Agreements could be worked out between federal agencies and local communities to make needed land available. This could take the form of land trades in which federal agencies exchanged land suitable for development for other land in the community. The National Forest Service has already established a precedent for this in its exchanges with some of the region's corporate landholders.

Absentee corporate ownership contributes to a substantial outflow of capital from the areas in which it is the dominant pattern of ownership. While local banks in such areas may appear to have tremendous assets, many of these assets are not money available for use in the community. Instead, it is on its way to banks in other areas of the country that are patronized by the parent corporation. Programs developed to induce these owners to invest their wealth in the local community might have limited success in increasing the availability of private local capital. Most likely, though, it is only in the public realm that significant amounts of new local capital can be created. The way that it can be created is through taxation on absentee corporate land that is sufficient to provide much of the local capital necessary for economic development.

Broadly defined, the infrastructure of a local community includes not only the available utilities (i.e., water, sewer, electricity) and transportation networks, but also facilities such as schools, hospitals, and parks. The maintenance of any of these at adequate levels seems to be more difficult in areas of concentrated corporate and/or absentee ownership. Both the lack of available land and the lack of local capital (whether public or private) contribute significantly to the problem. Fair and equitable taxation of the land and minerals, when combined with the possibility of local government condemnation of land for public use, could go a long way toward upgrading the infrastructures of many mountain localities.

Problems with both the quantity and quality of housing have long been chronic in the central Appalachian coal counties. What is less well known is that they have been severe in other areas of Appalachia as well. A major influence on this housing situation has been the direct and indirect effects of absentee corporate and government ownership. The direct impacts of such ownership patterns are: restrictions on the availability of land; barriers on financing where mineral rights are severed from surface ownership; and inflation of prices of land available to the local housing market. Indirect impacts include: lack of financing, inadequate provision of services such as water and sewage treatment, and competing land use patterns (e.g., stripmining).

The availability of land for housing has two aspects to it: whether there is any land available at all and whether any available land is affordable for the local population. In many central Appalachian coal counties there is virtually no land on the market. In some areas where there has been extensive resort development, there may be ample land available but at prices that are out of reach for much of the local population. This escalation of land prices has also been apparent in many coal counties. The two aspects of the housing problem may require somewhat different responses in terms of public policy.

In the first instance, the primary concern is to make land available where housing is needed, rather than the usual practice of providing housing only where there is land for sale. To facilitate this, state and local agencies should be empowered to condemn land for use in meeting local housing needs. This power is particularly needed in cases where absentee corporate owners do not willingly make such land available. In order to make the development of housing possible on any condemned land, such agencies would also need the power to condemn any land necessary for the development of needed services such as water and sewage treatment. Such condemnation procedures should be subject to public review to insure against their abuse.

The other aspect of the housing problem has to do with making it possible for local residents, whatever their economic circumstance, to purchase land and homes. In order to do this, local and regional capital reserves would be necessary. One option here might be the development of local, state, or regional land banks, initially capitalized by public funds and under perpetual public control. At the same time, many of the urban-oriented and unrealistic restrictions on publicly funded housing programs could be reevaluated in light of the realities in rural Appalachia. Also, in developing any housing strategy in Appalachia, the experiences of those private nonprofit groups that have worked on housing problems for years should be utilized (e.g., the member groups of the Federation of Appalachian Housing Enterprises).

The provision of housing may not mean much, however, unless there are sufficient restrictions on land use so as to prohibit contiguous development that is destructive of that housing. Of particular concern in central Appalachia is the possible effect of strip-mining. The homes of countless central Appalachians have been threatened and in some cases destroyed because the surface owner could not control the exploitation of the mineral under that surface. The establishment of the priority of surface owner's rights over those of the mineral owner would seem a necessary step in the protection of homeowners in areas like eastern Kentucky. Indeed, most counties in Appalachia could benefit from land use policies that protect and preserve badly needed housing.

If one can believe the promotions of regional officials and energy interests, Appalachia is likely to experience a sustained boom in energy development over the next two decades. Recent ownership and leasing trends do in fact indicate that the energy conglomerates and their subsidiaries anticipate a boom period. While the national and international economy and the politics of energy development may alter their plans somewhat, there can be no doubt that they see Appalachian energy resources as a good investment.

Beginning in the 1960s, large energy conglomerates (especially oil companies) have now gained control over much of the coal reserves in the region. This control is much greater than the ownership data itself would indicate, since the leasing of minerals is extensive and accounts for additional thousands of acres. The capital and technical resources of these corporations provide for the application of ever-larger scale technologies to the extraction of coal in the region.

The patterns of absentee ownership and control which are historically characteristic of central Appalachia have now extended outward to other counties. In the Eastern Overthrust Belt, there has been extensive speculation in oil and gas leasing by major oil companies. Speculation in and plans for the development of new minerals such as uranium and other strategic metals are now evident in some areas of Appalachia. These ownership trends and their associated potential development are proceeding with little comprehensive planning and practically no opportunity for citizen input. The possible impacts of this development on local communities demand more than that.

Leasing is a form of de facto ownership and indications are that leasing is becoming a primary strategy for the corporate control of the region's resources. However, residents in many affected areas are unfamiliar with mineral leasing arrangements and as a result are often at a disadvantage. Fraudulent leasing practices are not an uncommon occurrence, as recently illustrated in the leasing of oil shale in central Kentucky. In that situation the state had to threaten legal action before farmers were able to cancel fraudulent leases.

At a minimum, state and local officials as well as citizens' groups should monitor and publicize leasing activities in their areas. All such leasing activities should become public information through required reporting and recording at the time the lease is transacted. Further, eduational programs should be developed to inform landowners in those areas of their leasing rights and the potential impacts of energy development on their land and communities. Most often, corporations keep both their leasing activities and development plans concealed until it is too late for community residents to affect their decisions. Responsible action by county governing bodies could prevent these surprises. By the simple act of requiring public hearings when leasing activities begin in the county, for instance, they could insure that there was citizen awareness of the activity.

The accumulation of vast acreages of mineral rights and land by large energy conglomerates has produced a scale of strip-mining previously unknown in the mountains. Proposed strip-mine operations of several thousand acres are not unheard of. The negative effects of strip mining are well known as are the limited economic benefits for the localities involved. Yet, in spite of opposition by citizens' groups (e.g., Friends of the Little Kanawha in West Virginia), strip-mining continues apace. In Kentucky, repeated efforts to do away with the broad-form deed in both the courts and the legislature have met with little success. That deed essentially gives prior rights to the owners of minerals and allows them to extract those minerals with any method they desire. There has been progress elsewhere, though. The Tennessee Supreme Court, for instance, recently upheld a newly passed Surface Rights Act, which protects the rights of surface owners where the mineral rights have been severed. Kentucky would do well to follow that lead.

In light of the level of strip-mining now taking place in the region, it is ironic that the federal government has moved to weaken regulation of such activities. If anything, there should be closer regulation and more rigorous enforcement of state and federal regulations that protect the land and water from the impacts of strip-mining. The Office of Surface Mining should be strengthened, rather than reduced to ineffectiveness. With the primary responsibility for enforcement shifting to the states, a strong and vigilant federal agency is needed to insure that states comply with all regulatory provisions. Their records have not been that impressive in the past.

One of those provisions that has been under-utilized is the one referring to areas being designated as "unsuitable for strip mining." To have an area so designated has proven almost impossible. In the opinion of the Task Force, many more areas should have been declared unsuitable and thus preserved for more constructive land uses.

Over the past decade there has been a dramatic loss of farmland in Appalachia. Present ownership trends indicate that agricultural lands will experience increasing pressures from a number of sources: expanded energy development, damage from strip-mining, inflated prices, increasing property taxes, and the conversion of agricultural land to other uses. The decline in the regional agricultural economy is evident in the loss of acreage and number of farms, the low percentage of land devoted to agriculture, the percentage of farmers engaged in nonfarm occupations, and the increasing age of farmers.

The influence of ownership patterns here can be illustrated in several relationships. In the survey counties, the greater the corporate control of land, the lower the percentage of land devoted to agriculture. Absentee ownership and the concentration of ownership was also associated with low use of land for farming. The trends that are associated with these ownership patterns point to the demise of agriculture as a significant part of the regional economy unless appropriate action is taken to avert it.

The options for action are several here. One option would be to apply present use or agricultural assessments in all counties with agricultural lands. The present use assessment usually provides for a lower tax assessment as long as the land is maintained in its current use (i.e., agriculture). The farmer can usually take advantage of the lower assessment so long as he/she does not convert the land to other uses within a certain number of years (e.g., to resort or residential development). If conversion takes place sooner than that, the farmer must pay a higher rate for those years in which the present use assessment was taken. The intention is that the lower assessment will make it more likely that the farmer will keep the land in agricultural use rather than sell it for development or other nonagricultural uses. The experience with present use assessment suggests that it works only when accompanied by an educational program to insure that all eligible owners are informed about such assessments and how to use them.

Another vehicle used in several states is the agricultural assessment. Under this scheme land designated as agricultural is assessed at a much lower rate than other types of property. The definition of what is agricultural land is usually very restrictive. However, the intention often does not correspond to the actual practice. For instance, in Kentucky and Alabama large properties held for mining, not for agriculture, have often taken advantage of the agricultural assessment rates. In some eastern Kentucky counties, the property valuation administrators were routinely granting agricultural assessments to mineral properties above a certain acreage. So, if the agricultural assessment is to work as an instrument for the protection of agricultural land, its implementation must be as strict as its intention.

Energy development poses another threat to agricultural land in the mountains. Any energy development in agricultural areas should be undertaken only after extensive review of its impacts on farming, a review that includes maximum citizen input. The same could be said for recreational/tourist development. Educational programs are necessary to inform farmers of the possible impacts of both energy and tourist development. Some areas may need to be declared unsuitable for either because of the negative impacts they would have on prime agricultural areas. This protection could take either of two forms, among others. First, restrictions could be placed on the amount of farmland which could be held for nonfarm use (e.g., as in South Dakota). This would limit the amount of farmland that could be bought by large corporations or individuals to be held for speculative purposes. Second, restrictive zoning could be enacted that would protect farmland from the encroachment of energy development, resorts, and second-home development. For instance, this might involve zoning a section of the county as agricultural.

Even though the needs for housing, agriculture, economic development, and energy development involve competition among various land uses, systematic land use planning and regulation is virtually nonexistent in most rural counties of Appalachia. In this environment of little or no regulation, the land use decisions are made by the larger and more powerful owners. Such decisions are usually made in terms of their own interests and not of the needs of the majority of people in the community. In the case of large landholders, single decisions can affect entire areas, even though the affected public has had little or no say in the decision. In many cases, they may even have been unaware that a decision was pending until it was well on its way to implementation.

To alleviate this situation, land use mechanisms must be developed which insure broad-based citizen participation and which have the power to regulate land use in the interest of the larger community. Traditional zoning boards have fallen short here because they are usually dominated by special interest groups. So, land use boards should be developed which mandate the participation of a cross-section of the community's population. One model for such a board is that of the local public utility, a model that would assure public control of land resources. Such a board would be empowered to purchase land and preserve it in the public interest. While still allowing for private property and traditional land use control, this model also provides for local public ownership that could relieve patterns of absenteeism and concentrated ownership.[6]

All of the strategies for change discussed in the preceding pages can be very significant in the extent to which they improve the quality of life in Appalachian communities. However, few of them strike directly at the root of the problem—the corporate domination of Appalachia that rests on the concentrated ownership of the region's land and other natural resources. Most of them instead attempt to lessen the negative consequences of that problem. Yet, struggles around these issues are no less important as a result. To force a corporate mineral owner to pay its fair share of property taxes is certainly no small feat. However, until the structure of corporate domination of Appalachia's resources is radically altered, citizens of the region will endlessly be struggling to lessen the negative consequences. What is needed is a change in the ownership patterns themselves.

Somehow actions must be taken which deal with the underlying problems of concentrated and absentee ownership of land and mineral resources. Mechanisms have to be found by which people of the region can gain more access to, control over, and benefit from the land and its resources. What this implies is some measure of land reform in the region. Options for land reform which protect and benefit all Appalachian communities and their inhabitants can and should be developed. Possible options range from the use of eminent domain for meeting community needs to programs for limiting excessive corporate ownership of land for speculative purposes: from developing broad, new programs for land redistribution to broader public ownership and control of the land and resources.

For too long there has been a pervasive myth that land reform is only needed in countries of the Third World, ignoring the urgent need for land reform in the rural areas of this country. Nowhere is the need for such reform more obvious than in Appalachia. The negative impacts of concentrated corporate ownership in the region are too well documented to be ignored. It is now long past time for public discussion of land reform options in the region. The future of Appalachia and its people is too important to do otherwise.

Ultimately, the measure of the landownership study will not be how many people read it, or what acclaim it receives, but how many citizens of Appalachia use its findings. In that respect it has had an exciting history since its release in the early spring of 1981. It has been used to filibuster against tax relief for the timber industry in the Alabama state legislature. In Tennessee, Save Our Cumberland Mountains (a grassroots group of coalfield residents) has organized several county tax efforts around the study's findings. In West Virginia, it was introduced as evidence in what turned out to be a monumental court case in which the court ruled that rural areas in the state were unconstitutionally discriminated against due to the inadequacies of the property tax system used to fund their schools.[7] In Kentucky, the Kentucky Fair Tax Coalition used its findings to spearhead a campaign for a state tax on unmined minerals as well as to bring local challenges to corporate assessments. In Virginia, it became the basis for local tax challenges and the impetus for attempts to form a Virginia Land Alliance. And finally, in North Carolina, it gave rise to a minerals leasing conference and local educational efforts around such leasing.

Viewed as part of the ongoing struggle for social justice in the mountains, much of the story of the landownership study has yet to be told. The remainder of the story will unfold gradually as its findings continue to be used by citizens' groups in the region in their various struggles against the impacts of concentrated corporate and absentee ownership. It will also be told as some vision of an Appalachia free of that domination begins to develop in the course of their efforts.

APPENDIX ONE

Fifty Top Owners and Other Data

Table A-1. Definitions of Types of Counties

County Type	Number of Counties	Definition	Source of Data
High coal	42	Known coal reserves greater than 100 million tons*	*Atlas of Environmental and Natural Resources in Appalachia*, ARC, 1977
Medium coal	16	Known coal reserves less than 100 million but greater than zero	
No coal	22	No known reserves	
High agriculture	33	Total value of sales greater than $5 million.	Census of Agriculture, 1974
Low agriculture	47	Total value of sales less than $5 million	
High tourism	19	Percentage of service receipts in tourism and recreation greater than 24.4%**	Census of Selected Service Industries, 1972
Low tourism	52	Percentage of service receipts in tourism and recreation less than 24.4%	

*Refers to bituminous and semianthracite coal resources remaining in the ground as of 1 January 1973.
**Includes percentage of service receipts in hotels, motels, trailer parks, and campgrounds, plus percentage in amusements and recreation. Counties with both variables missing were excluded.

Appendix One

Table A–2. Fifty Top Surface Owners in Eighty Appalachian Counties

Name and Headquarters	Principal Business	Type of Company	Total Surface Acres	Chief Location of Holdings
1. J. M. Huber Corp., Rumson, N.J.	diversified, timber and wood	private	226,805	Tenn., Ky.
2. Bowaters Corp. (Hiwassee Land Co.), London, Eng.	wood products	public	218,561	Tenn.
3. N&W Railroad (Pocahontas Land & Pocahontas-Ky.)[a], Roanoke, Va.	railroad, transportation	public	178,481	W.Va., Ky., Va.
4. Koppers Co., Pittsburgh, Pa.	diversified chemicals, metals, coal gasification	public	169,796	Tenn.
5. U.S. Steel, Pittsburgh, Pa.	steel	public	168,911	Ala., Ky., Tenn., W.Va.
6. Georgia-Pacific, Atlanta, Ga.	wood products	public	139,441	W.Va., Va., Ky.
7. Pittston Corp., New York, N.Y.	coal	public	137,650	Va.
8. Tenneco, Inc. (Tennessee River, Paper and Pulp), Houston, Tex.	oil, land, packaging	public	98,751	Ala.
9. Continental Oil (Consolidated Coal), Stamford, Conn.	oil, gas, petrochemicals, coal	public	84,403	W.Va., Va., Ky.
10. Gulf States, Tuscaloosa, Ala.	paper and wood products	public	78,054	Ala.
11. Chessie System, Inc. (Western Pocahontas, C&O Railroad), Baltimore, Md.	holding company, transportation, petrochemical	public	76,805	Ky., W.Va.
12. Weyerhauser, Seattle, Wash.	wood products	public	65,005	Ala.
13. Coal Creek Mining & Mfg., Knoxville, Tenn.	coal lands	private	64,374	Tenn.
14. Champion International, Stamford, Conn.	building materials, paper, furniture	public	63,405	Ala., N.C.
15. Penn Virginia Corp., Philadelphia, Pa.	coal lands	public	62,893	Va.
16. Berwind Land Co. (Kentland Co.), Philadelphia, Pa.	coal and natural resources; other diversified products	private	60,881	W.Va., Ky., Va.

Appendix One

17. Kentucky River Coal, Lexington, Ky.	coal lands	private	56,279	Ky.
18. Bethlehem Steel, Bethlehem, Pa.	steel, steel products	public	47,132	Ky., W.Va.
19. Mead Corporation (Georgia Kraft Co.), Atlanta, Ga.	paper and wood products	public	46,765	Ala.
20. Rowland Land Co., Charleston, W.Va.	coal lands	private	44,867	W.Va.
21. Bruno Gernt Estate, Allardt, Tenn.	coal, timber	private	42,317	Tenn.
22. Union Carbide, New York, N.Y.	chemicals, carbon products	public	41,060	W.Va.
23. Brimstone Co., Dover, Del.	coal lands	private	40,261	Tenn.
24. Soterra, Inc., Delaware, Ohio	unknown	private	39,917	Ala.
25. Stearns Coal and Lumber, Stearns, Ky.	coal land, timber	private	38,934	Tenn.
26. Southern Co. (Alabama Power), Atlanta, Ga.	utility	public	38,736	Ala.
27. Plateau Properties, Crossville, Tenn.	land, mining	private	38,430	Tenn.
28. Lykes Resources, Inc. (Youngstown Mine), Pittsburgh, Pa.	steel	public	38,071	W.Va., Va.
29. Alabama By-Products, Birmingham, Ala.	coal, coke, chemicals	public	34,365	Ala.
30. American Natural Resources (Virginia Iron Coal & Coke) Detroit, Mich.	gas, coal	public	33,155	Va., Ky.
31. Beaver Coal Co., Beckley, W.Va.	coal lands	private	32,994	W.Va.
32. St. Joe's Minerals (Tennessee Consolidated Coal). Jasper, Tenn.	coal, other minerals	public	32,323	Tenn.
33. Hugh D. Faust, Knoxville, Tenn.	coal lands, timber	individual	32,021	Tenn.
34. Jim Walter Corp., Birmingham, Ala.	pipe, metals, coal, building materials	public	31,721	Ala.
35. Dingess Rum Coal Co., Huntington, W.Va.	coal lands	private	31,282	W.Va.

Table A-2. Continued)

Name and Headquarters	Principal Business	Type of Company	Total Surface Acres	Chief Location of Holdings
36. Crescent Land Co., Charlotte, N.C.	land development	private	31,200	N.C.
37. Carolina Rite Co., Miami, Fla.	timber/pulp	private	30,330	N.C.
38. Mower Lumber, New York, N.Y.	timber, coal lands	private	29,792	W.Va.
39. Cole Interests, Huntington, W.Va.	coal lands	private	27,385	W.Va.
40. Albert Holman, Tuscaloosa, Ala.	coal lands	individual	26,284	Ala.
41. Kentenia Corp., Boston, Mass.	coal lands	private	25,335	Ky.
42. Cotiga Development Co., Philadelphia, Pa.	coal lands	private	25,081	W.Va.
43. Eastern Gas & Fuel Co. (Eastern Associated Coal), Boston, Mass.	coal, coke, gas	public	24,516	W.Va.
44. American Electric Power (Franklin Real Estate), New York, N.Y.	utility	public	22,775	Va., Ky.
45. Blue Diamond Coal Co., Knoxville, Tenn.	coal, land	private	22,206	Tenn.
46. Eastern Property Trading Co., Atlanta, Ga.	real estate	private	22,120	Ala.
47. Quaker State Oil (Kanawha Hocking and Valley Camp Coal), Oil City, Pa.	oil	public	21,175	W.Va.
48. Wilson Wyatt, Louisville, Tenn.	attorney	individual	21,131	Tenn.
49. Grandview Mining Co., Chattanooga, Tenn.	coal, land	private	21,116	Tenn.
50. National Steel, Pittsburgh, Pa.	steel	public	21,000	Ky.
Total			3,006,322	

Source: Appalachian Land Ownership Study, 1980.
aMerged with Southern Railway after completion of study.

Appendix One

Table A-3. Fifty Top Mineral Owners in Eighty Appalachian Counties

Name and Headquarters	Principal Business	Type of Company	Total Mineral Acres	Chief Location of Holdings
1. Columbia Gas System, Wilmington, Del.	natural gas, holding company	public	342,236	W.Va.
2. N&W Railroad* (Pocahontas Land & Pocahontas-Ky.), Roanoke, Va.	railroad, transportation	public	201,950	Ky., W.Va.
3. Continental Oil (Consolidation Coal), Stamford, Conn.	oil, gas, petrochemicals, coal	public	193,061	W.Va., Ky.
4. Pittston Corp., New York, N.Y.	coal	public	185,254	Va.
5. Occidental Petroleum (Island Creek Coal), Los Angeles, Ca.	gas, oil, petrochemicals, coal	public	144,741	W.Va., Ky., Va.
6. Berwind Land Co. (Kentland Co.), Philadelphia, Pa.	coal, natural resources	private	108,561	Ky.
7. American Natural Resources (Virginia Iron Coal & Coke), Detroit, Mich.	gas, coal	public	80,705	Va.
8. U.S. Steel, Pittsburgh, Pa.	steel	public	71,601	Ala., Tenn., W.Va.
9. Republic Steel, Cleveland, Ohio	steel	public	67,252	Ala.
10. Georgia-Pacific, Atlanta, Ga.	wood products	public	67,027	W.Va.
11. First National Bank of Birmingham, Birmingham, Ala.	bank, holding company	private	66,991	Ala.
12. Diamond Shamrock (Falcon Seaboard), Cleveland, Ohio	oil, gas, chemicals, coal	public	66,928	Ky.
13. Deep Water Properties (held through First National Bank of Birmingham), Birmingham, Ala.	financial trust	private	66,038	Ala.
14. Cherokee Mining, Houston, Tex.	coal	private	60,294	Ala.
15. National Steel, Pittsburgh, Pa.	steel	public	60,000	Ky.
16. Reynolds Metals (Reynolds Minerals), Richmond, Va.	ore, chemicals, aluminum	public	58,000	N.C.

Appendix One

Table A-3. Continued

Name and Headquarters	Principal Business	Type of Company	Total Mineral Acres	Chief Location of Holdings
17. Wilson and Maryanne Wyatt, Louisville, Ky.	attorney	individual	57,614	Tenn.
18. Chessie System, Inc. (Western Pocahontas, C&O Railroad), Baltimore, Md.	holding company, transportation, chemicals	public	56,830	W.Va., Ky.
19. Rowland Land Co., Charleston, W.Va.	coal lands	private	54,474	W.Va.
20. North Alabama Mineral Division Co., no address	minerals	unknown	50,141	Ala.
21. J. M. Huber, Rumson, N.J.	diversified, timber and wood products	public	47,759	Tenn.
22. Quaker State Oil (Kanawha Hocking and Valley Camp Coal) Oil City, Pa.	Oil	public	47,711	W.Va.
23. Wesley West, Houston, Tex.	coal lands	individual	46,682	Ala.
24. Beaver Coal Co., Beckley, W.Va.	coal lands	private	44,807	W.Va.
25. Plateau Properties, Crossville, Tenn.	land, mining	private	42,038	Tenn.
26. Union Carbide, New York, N.Y.	chemicals, carbon products	public	41,689	W.Va.
27. Alabama By-Products, Birmingham, Ala.	coal, coke, chemicals	public	41,001	Ala.
28. Charleston National Bank, Charleston, W.Va.	bank, holding	private	40,566	W.Va.
29. Cotiga Development Co., Philadelphia, Pa.	coal lands	private	39,648	W.Va.
30. Mower Lumber, New York, N.Y.	timber, coal lands	private	36,776	W.Va.
31. Eastern Gas & Fuel Co. (Eastern Associated Coal), Boston, Mass.	coal, coke, gas	public	35,066	W.Va.
32. Sun Oil (Shamrock Coal), Radnor, Pa.	oil company	public	34,927	W.Va.
33. Southern Railway*, Washington, D.C.	rail transport	public	34,877	Ala.

Appendix One

34.	Coal Creek Mining & Mfg., Knoxville, Tenn.	coal lands	private	34,042	Tenn.
35.	Lykes Resources, Inc. (Youngstown Mine), Pittsburgh, Pa.	steel	public	33,972	W.Va.
36.	L&N Railroad, Lexington, Ky.	railroad	public	32,575	Ala.
37.	Penn Virginia Corp., Philadelphia, Pa.	coal lands	public	32,267	Va.
38.	Dayton Hale, Tuscaloosa, Ala.	banker, real estate	individual	31,600	Ala.
39.	Julius Doochin, Nashville, Tenn.	contractor, coal lands	individual	31,000	Tenn.
40.	Dingess Rum Coal Co., Huntington, W.Va.	coal lands	private	30,186	W.Va.
41.	Neva McMullen, Washington, N.C.	coal lands	individual	29,901	W.Va.
42.	Drummond Coal Co., Jasper, Ala.	coal mining & coal lands	private	29,038	Ala.
43.	W. R. Burt, Lexington, Ky.	coal, land	individual	28,701	Ala.
44.	Bruno Gernt Estate, Allardt, Tenn.	coal, timber	family	28,354	Tenn.
45.	Cole Interests, Huntington, W.Va.	coal lands	private	28,046	W.Va.
46.	Southern Land and Exploration, Tuscaloosa, Ala.	coal lands	private	27,284	Ala.
47.	Consolidated Goldfields (Goldfield Mining Corp.), London, Eng.	multinational mining interests, including South Africa	public	26,706	Tenn.
48.	National Shamuts Bank of Boston, Boston, Mass.	bank, holding	private	26,453	Va.
49.	Kentucky River Coal, Lexington, Ky.	coal lands	public	26,272	Ky.
50.	Hagan Estate, Tazewell, Va.	coal, land	individual	25,854	Va.
Total				3,095,496	

Source: Appalachian Land Ownership Study, 1980.
*Merged with Southern Railway after completion of study.

Table A-4. Summary of Aggregate Data Collected for Eighty Appalachian Counties

Type of Data	Source	Year
Land-Use Data		
Known coal reserves	ARC* Atlas	1973
Known coal production	Ohio River Basin Energy Study	1977
Known agricultural data	Census of Agriculture	1974
Economic Data		
Income characteristics	ARC Data Bank	1970, 1974
Labor-force characteristics	ARC Data Bank	1970, 1977
Employment characteristics	ARC Data Bank	1970, 1977
Banking-deposit characteristics	City County Data Book	1976
Industry characteristics	Census of Manufacturing,	1972
	Census of Mining	1972
Agricultural characteristics	Census of Selected Service	1972
	Industries, Census of Agriculture	1969, 1974
Community Data		
Migration and population characteristics	ARC Data Bank	1970, 1975
Housing characteristics	Census of Housing	1970
Health characteristics	ARC Data Bank	1974
Education characteristics	Census of Population	1970
Fiscal Data		
County revenue sources	Census of Governments	1977
County budget expenditures	Census of Governments	1977
County property taxes	Census of Governments	1977

*Appalachian Regional Commission.

APPENDIX TWO

Methodology of the Land Study

The study of landownership patterns is difficult from a methodological point of view. Both the availability and the quality of data leave much to be desired. Generally speaking, there are few repositories of such information other than at the county level. Even at that level there is not enough standardization across counties within the same state, much less across state lines, to allow for easy comparability of ownership data. Public and nonprofit private ownership data are often not even recorded at the local level. State property and taxation laws give local officials much leeway in their implementation, while offering insufficient incentives for accurate and up-to-date record-keeping. In the Appalachian region the severance of mineral from surface ownership and the complex interlocks between corporations that own and/or mine mineral lands further compound the picture. The resulting property ownership records are often confusing patchworks of contradictory and incomplete data.

Previous landownership studies in Illinois, Ohio, Kentucky, North Carolina, Tennessee, West Virginia, and other states had demonstrated that reasonably accurate data could be obtained in spite of these difficulties. To one degree or another, however, most of them shared several limitations that the Appalachian Landownership Task Force wished to transcend: (a) restrictive geographical scope and lack of interstate comparability; (b) lack of substantive county case studies and in-depth investigation of impacts; and (c) the failure to train and utilize area residents as researchers. What most of the studies shared in common was the utilization of county property and tax records as basic sources of information. It was with respect to courthouse research (how to do it and what to look for) that previous studies were most useful.

One instructive study had been undertaken by the Illinois South Project in the coalfields of southern Illinois in an attempt to determine who owned the minerals there. Ownership was defined by the individual or company paying the tax bill on a particular parcel of land. For the basic research, the project used the Supervisor of Assessment Office Books found in the county courthouse. Other county resources, such as the Tax Collectors Books and the Grantor and Grantee Books in the Recorder of Deeds Office, were consulted when necessary. Major problems encountered were the result of the quality of data and the rapid turnover of mineral ownership in some counties.[1]

Another recent study whose methodology influenced our own was conducted in a five-county area of southeastern Ohio. Data for the study was taken from the county tax records made available through the offices of the treasurer and auditor. Information collected included name and address of the owner, location of the property, acreage, type of ownership, assessments, and so on for every piece of property in the county, a research task that was very time consuming. Owners were classified as nonresident, corporate, or public, according to the title ownership listed in the county tax books. A second part of the study included mailing questionnaires to a sample population of nonresident owners to determine more specific information on those owners and their property. Additional time was spent mapping absentee-owned property in each county.[2]

During the late 1960s Richard Kirby did one of the first ownership studies in eastern Kentucky for the Appalachian Volunteers. In an attempt to answer the question "Who owns east Kentucky?" he went to the tax books in eleven east Kentucky courthouses.[3] In 1977, the University of Kentucky investigated ownership in two eastern Kentucky coal counties, Harlan and Perry. In each county, owners of over 100 acres were obtained from the Tax Assessor's books, along with the applicable property assessments. Owners were also typed according to ownership (corporate or individual) and residence (local or absentee).[4]

In the summer of 1974, the Mountain Land Use Project of the North Carolina PIRG spent four months collecting landownership data and conducting interviews with local officials and landowners in the western part of the state. In selecting counties to be studied, they first profiled all twenty-four mountain counties on the basis of certain general characteristics of population, geography, and apparent development. The resulting list of counties was broadly representative of the total western Carolina area. Researchers examined the county tax records, listing the parcel, its acreage, and the address of the owner. In each county the information was analyzed for 1968 and 1973, in order to assess changes over the five-year period. Data were then analyzed by local and nonlocal holdings and by size of tracts. Only limited efforts were made to trace parent companies in the case of corporate ownership.[5]

Earlier, in 1971, three Vanderbilt students had investigated landownership and taxation patterns in five major eastern Tennessee coal-producing counties. Tax rolls in the counties were used as the basic data source, while information on coal leasing was found in County Deeds and Records Offices. Further data on rates of coal production were obtained from the Bureau of Mines, Tennessee Department of Geology, Tennessee Department of Labor, and the Tennessee Department of Conservation. Corporate profiles were developed from standard financial sources and from interviews with corporate officials.[6]

Also, in 1971, a major study of landownership and taxation patterns was conducted in fourteen West Virginia coal-producing counties. Data was collected from copies of county land books filed in the state land office. From the county-by-county ownership data, listings of large landowners were then developed.[7] A few years later a journalist with the *Huntington Herald* further investigated the ownership of land and minerals in West Virginia, pointing out that the control of these resources was extended through the leasing process.[8]

While these studies demonstrated the feasibility of landownership research through the use of courthouse records, none of them reflected an integrated research strategy involving the citizens affected by that ownership. The landownership study was viewed as a decentralized, cooperative research process involving area citizens in the definition of the research problem, development of the research instruments, collection of the data, and use of that data (i.e., education and action around the findings of the study). If we may think of methodology in a broader sense than is usually the case, the study was guided by the methodology of participatory research, an integrated strategy for citizen utilization of research to solve the societal problems affecting them.[9] Much of the following discussion focuses on this process of research rather than either the techniques of research or analysis. As such, it speaks as much to the control of research as to the specific techniques of research. Ultimately, the control of research is the more important methodological question.

The proposal to undertake the study arose out of the Appalachian Land Ownership Task Force, a coalition of scholars, citizens' groups, and individual citizens affiliated with the Appalachian Alliance. Several of the academic members belonged to the Appalachian Studies Conference, an organization of scholars formed in 1978 to further research and understanding of the region. In addition to the provision of comprehensive information on landownership, the project was also to serve as a model by which local residents and citizens' groups could investigate local and regional issues. It was this citizen participation in the research process that was responsible for the unique methodology of the landownership study.

Planning for the landownership study took place over several months in the fall of 1978 and the winter and spring of 1979. Over these months members of the Task Force met to formulate the goals, methodology, and structure of the proposed study. After reviewing the methods of previous landownership studies, they developed a research strategy suitable for gathering accurate information through the involvement of local citizens in the research process.

The Task Force assumed responsibility for the overall coordination of the research project. Members were responsible for the recruitment of regional and state staff who would in turn assume responsibility for the day-to-day coordination of the research. State task forces, made up of citizen representatives from each of the states in the study area, were also formed. These groups of citizens were instrumental in the selection of state coordinators and other aspects of the study such as the selection of sample counties and identification of critical impact areas.

Training field researchers and developing a suitable research instrument were part of the same task, since area citizens were to participate in every phase of the research process. They were involved in the development of the various research techniques used in the course of the study. Their insights, along with those of resource people with experience in landownership research, were critical to the planning of the first and subsequent training sessions. The first workshop in May 1979 brought together some fifty people to prepare for the field research. Later training sessions were held at the state level during the following summer to deal with research problems that were state specific.

The first workshop had to deal with several challenges. Training was necessary in the types of resources available in the courthouse, development of rapport with office personnel, how to trace down the real owners, problems likely to be encountered, and so on. A coding sheet had to be prepared with which field researchers could record the ownership information found on the tax books. Workshop participants also had to become familiar with other resources that would help them to identify connections between corporate owners. Training was also needed in the research necessary to do county level case studies.

Most of the relevant information about landownership and taxation is available in the county courthouse, and public access to it is, in most cases, protected by state law. Most county officials and their staffs respect this right and are cooperative. In those few instances where they are not, a reminder of citizen rights to the data may be necessary. The primary source of landownership information to which the researcher wants access is the property tax books. These are compiled yearly and are normally found in the office of the county tax assessor or other county official responsible for sending out tax notices.

Two separate books record tax information on (a) personal property, including vehicles, machinery, livestock and, in some cases, leases; and (b) real property, including land, minerals, and buildings. These books are divided into sections according to voting or magisterial districts within the county. If the precise boundaries of those districts are not known, the researcher can usually identify them by looking at an official county map.

Property owners are listed alphabetically within each district, although, in some cases, commercial (corporate) property is listed separately at the end of each district. So, if no corporate listings are apparent at first, or if a particular corporate holding seems absent, the researcher should check to see if there is a separate listing. Sometimes publicly owned land and other tax-exempt land is also listed at the end of each district's tax rolls.

The real property tax books will usually show some or all of the following: the owner's name and address; a description of the property; the number of acres or size of the lot; the type of minerals present, if any; the property's location; the value of the land, minerals, and improvements; the amount of tax billed to the owner; the tax map parcel number; and the deed book reference. If the property is designated as fee or fee simple, ownership includes both the land surface and minerals underneath it.

Appendix Two

Most of the items found on the property tax rolls are self-explanatory, but a few may need elaboration:

1. Every piece of property listed should have either an assessed or an appraised value next to it. "Appraised value" usually refers to a supposed calculation of "true and actual" value or "market" value. In many cases, appraised values are outdated and/or considerably lower than true value. "Assessed value" is usually a fixed percentage of the appraised value, used to determine the amount of tax due. The percentage set for a particular class of property should be uniform throughout the county. If only the assessed value is listed, one can compute the appraised value of a piece of property by knowing the fixed percentage.
2. The actual tax billed is figured by multiplying the county's tax rate times the assessed value. For example, in a county where the tax rate is $3 for every $100 of property value, the tax on a piece of property appraised at $6,000 would be: ($6,000÷$100) X $3 = $180. In comparing taxes charged to different owners, the researcher must remember that there are usually different tax rates for different kinds of property. Commercial property, for example, is often taxed at a higher rate than agricultural property. The tax assessor's staff can usually explain these rate differences.
3. The parcel number refers to the county tax maps and can be used to determine the exact location of a particular piece of property by matching the number to its place on the map. The map is generally found in the assessor's office, but not all counties have them.
4. The deed book reference usually gives a volume and page number indicating the location in the county deed books where the deed and any leases on the property are listed.

Some counties also include recapitulations at the end of each district list or at the end of the volume itself. These can be useful in comparing the total assessed values and/or taxes paid by type of property, though much depends on how the totals are summarized.

Though property tax books were the primary source of information for this study, deed/lease books were often used to clarify ownership questions. Deed/lease books are especially helpful in tracing the history of corporate ownership of particular parcels and current leasing trends. These books are usually kept in the office of the county clerk or the county recorder. Sometimes deeds and leases are recorded in the same volumes, and sometimes they are listed separately. In either case, the volumes generally have a separate index where the names of all parties to property transactions are listed alphabetically each time a transaction occurs. Deeds are indexed according to both the "grantor" (the individual or company selling the property) and the "grantee" (the individual or company buying it). Indexes for leases are indexed by "lessor" (the individual or company leasing out the property) and "lessee" (the individual or company assuming the lease).

The indices refer to a volume and page number in the deed/lease books where the wording of actual terms of the deeds or leases can be found. In the case of leases, at least the following will be recorded: names of the lessor and lessee, the acreage involved, the length of time involved, the types of minerals, the royalties to be paid, and other conditions of the lease. In the case of deeds, the researcher will find: the buyer and seller, the date of the property transfer, the parcel's location and description, the terms and any special conditions of the transfer and, in many cases, the price paid and the total acres of land or minerals involved. Where leases and deeds are recorded in the same volumes, leases are usually appended to the deed and can be found by referring to the property through the grantor/grantee index.

Deed/lease books are important if the researcher wants to determine how much leasing is going on in the county, what kind and between what parties or they can be used to determine whether a particular company has leased in the county, how much it has leased and from whom. Suppose there is a rumor that Gulf Oil Corporation has been leasing in a given county. The researcher can go to the lessor/lessee index, look up Gulf Oil and determine if and when it has leased there. The index will refer to the volume and page where the actual terms of the lease can be found.

Who really owns or leases a particular piece of property may not be evident in the courthouse documents. This is particularly true in the case of corporate ownership, where the listed owner may only be a company subsidiary, regional office, or a front to conceal such ownership. So, the researcher may need to go elsewhere to discover true and actual ownership. For instance, the office of the secretary of state, usually in the state capital, records information on corporations that do business in the state. By letter or personal visit, any citizen can learn who the incorporators of a company are, where it was incorporated, who is on the board of directors, who are the current officers and some of the history of the company. Reports filed with the Securities and Exchange Commission in Washington by corporations of a certain type and size also reveal valuable information about subsidiaries, assets, and ownership. The researcher should ask for Forms 10-K and 8-K filed by the company in which he/she is interested.

Publications that may prove useful include: the *Keystone Coal Industry Manual, Moody's Industrial Manual, Standard and Poor's Register of Corporations, Directors and Executives,* and *Who Owns Whom* directories. These sources provide information on corporate histories, officials and directors, and relationships between subsidiaries and their parent companies.[10]

In selecting counties for the survey, the Task Force decided to include as many as possible. Selection of these counties was based on two criteria: (a) representativeness of the various types of landownership and use patterns in the region; and (b) the existence of local citizen interest in develop-

Appendix Two

ing, completing, and using the study. On the one hand the Task Force wanted a selection of counties that represented coal, agricultural, and recreational areas of the region. Previous studies and the divergent historical development of these areas led them to expect different ownership patterns and impacts. On the other hand, the Task Force wanted to insure the participation of local citizens in all phases of the research process. Given these two basic considerations, final selection of sample counties was made by the state task forces in cooperation with each state coordinator.

Using these criteria for county selection, eighty counties were chosen for the survey phase of the ownership study. The state-by-state breakdown gives the following number of counties in each state: Alabama (15); Kentucky (12); North Carolina (12); Tennessee (14); Virginia (12); and West Virginia (15). The original intention was to survey only seventy-two counties, but citizen interest led the Task Force to include eight additional counties. The percentage of Appalachian counties in each state included in the survey ranged from twenty-five percent in Kentucky to fifty-seven percent in Virginia. The eighty counties represented thirty-four percent of all the Appalachian counties in the six-state area.

In its attempt to develop a coding instrument that would allow for recording accurate and comparable ownership data across the eighty counties, the Task Force turned to other landownership studies for guidance. The southeastern Ohio study conducted by Nancy Bain and her associates was particularly helpful. She served as a consultant in the planning of the Appalachian study and provided the Task Force with coding instruments used in her study. During the May 1979 training workshop, participants used a revised draft of that coding sheet as the base from which to develop one suitable to the variations expected in a regional study.

A primary concern was to decide what property was to be coded. In the Ohio study every piece of property in each county had been recorded, a task that seemed unnecessary, if not impossible, in an eighty-county study. The Task Force decision was to record all owners of property in excess of 250 acres and all corporate or absentee owners holding twenty or more acres. Given the objectives of the study and the constraints of time and resources, such cut-offs made sense in light of what was known about landownership in Appalachia.

Previous studies of landownership in Appalachia had identified absentee and/or corporate ownership as major problems of the region. In the coal counties this was usually corporate ownership, whereas in recreational counties it was a combination of corporate, federal and, individual ownership. It was thus deemed important to identify as much of the absentee ownership as possible and all large holdings. The 250-acre limit insured the inclusion of the owners of large tracts, while the twenty-acre limit took into account most of the expected absentee ownership, even that in relatively small parcels. The net effect of numerous such parcels is often similar to that of larger absentee holdings. In case where it became evident that a single absentee or corporate owner held numerous parcels less than twenty acres, but which together might total twenty acres or more, the acreages were recorded and summed. Even with this added procedure, the twenty-acre limit probably led to under-reporting absentee individual ownership.

One further criterion was used in selecting the parcels to be coded. Researchers were instructed to code only that property lying in rural, unincorporated areas of the counties. It was in those areas where the Task Force expected to find concentrations of the types of ownership previously identified as contributing to the region's development problems. Their primary concern was to determine how those patterns affected the overall development of the counties and the region. The acreage limits used would have probably excluded most incorporated areas anyway, but parcels within such areas were not recorded regardless of size.

Using these guidelines, researchers identified land parcels as to their location in a state, a county, and a district within a county. Ownership of these parcels was then coded into four categories: individual, corporate, public, or private nonprofit. Ownership was determined by the name of the owner listed on the tax rolls or in other sources where necessary. Individual ownership was defined as ownership by one or more persons who did not constitute a business, level of government, or nonprofit organization. Corporate ownership referred to ownership by one or more persons who constituted a business organization. Public ownership was defined as ownership by either local, state, or federal government. Private nonprofit ownership referred to ownership by one or more persons who, for purposes of taxation, were classified as a nonprofit organization (e.g., a church or a college).

The diversity of tax record systems in the various counties and states posed some problems in determining ownership. Some ownership is simply not recorded in the tax assessor's office in many counties. For instance, public ownership is often not available, evidently because it is not subject to assessment and taxation. The amount of such ownership is usually available, however, from the appropriate federal or state agencies (e.g., National Forest Service, Department of Game and Inland Fisheries). These ownership figures can sometimes be obtained from the local planning district, but may not be as precise as needed. Private nonprofit ownership is usually not recorded either and, if not, there are no official sources to which the researcher can go to obtain the information. Some states, though, are now requiring that the county tax assessor keep a record of these types of property.

Two problems were evident in the recording of corporate ownership. First, land and minerals owned by utilities are usually not recorded on the county tax assessor's books. Where it is recorded, the information available is usually only partial, often with no notation of the exact acreage or taxes paid. Instead such information and the responsibility for taxation of utilities belongs to some state agency such as the Bureau of Public Works in West Virginia or the State Department of Revenue in Alabama. In some states this information is considered confidential and thus not available for public inspection. In those states where this is the case, corporate ownership may appear less than it actually is.

Appendix Two

Second, there is the problem of determining actual ownership. In many instances, the corporate owner listed on the tax books is either a subsidiary or regional office of some other corporate entity. Where corporate interlocks were well known, there was little problem in identifying the parent company (the real owner). In other cases such ownership was more difficult to trace. However, standard published sources as *Standard and Poor's, Who Owns Whom,* or *Moody's* yielded information on most of the corporate interlocks.

Extensive leasing in the mountains also confuses the ownership picture. Since active leases are often not listed in the same books as property ownership and in some instances seem hardly to be recorded at all, it was impossible in this study to record precisely the extent of leasing. However, interviews with local tax assessors, summary reviews of deed/lease books, and the county case studies indicated an acceleration in leasing activity in much of Appalachia (e.g., central West Virginia, northern Alabama, southwestern Virginia). If a thorough survey of leasing were undertaken, the result would probably demonstrate a much greater extent of absentee corporate control of mineral resources in the region that does the study of ownership alone.

Residence of owner is important for at least two reasons: (a) residence of the owner affects the use to which land is put and thus may have very different impacts on the local community;[11] (b) literature on land issues and previous studies in Appalachia point to absentee ownership as a key problem.[12] In this study the determination of residence of the owner was initially made on the basis of the address recorded in the county tax books. The residence of the owner was coded as either: in-county, out-of-county/in-state, or out-of-state. Problems in utilizing these coding categories resulted both from the categories themselves and from the variations in county record-keeping systems.

The problems in West Virginia illustrate the latter difficulty. In that state the tax assessor is charged with assessing property and preparing the tax books, while the county sheriff is responsible for sending out the tax bills. As a result, addresses of owners were not always available in the assessor's office, but rather in the sheriff's office. Researchers who were expecting the tax books to have such information had difficulty making an initial identification of residence. In those cases, records at the sheriff's office, state computer banks, and county phone books had to be used to identify residence. In the case of corporate owners, determination of residence could be made from a variety of available sources. Still, not all owners could be coded according to residence. Thus, both the absolute number of parcels coded and the number of acres coded for West Virginia are smaller in relative terms than in the other states surveyed. Also, the percentages of out-of-county and out-of-state holdings appear smaller than they actually are relative to other states. This is particularly true of absentee individual ownership.

A second problem was inherent in the definition of the residence codes. This was particularly so with the in-county classification. Postal routes often cross county and state lines, making it difficult to determine whether some addresses are in-county or out-of-county. Where counties border on other states, an out-of-state address may actually be in an adjacent county. In the latter case, the owner's significant attachments may be with the county in the adjacent state rather than with his/her own state of residence. It might be advisable for those undertaking future landownership studies to consider broadening the definition of local residence to include contiguous counties.

Total surface acreage was recorded for all surface land falling within our acreage limitations. The actual total acres for each of these parcels was usually available in the 1978 tax books used in the study. The exceptions were some public acreages, utility-owned holdings, and most private nonprofit acreages. Although supplementary sources were used to obtain as many of these as possible, the landownership study data under-records these types of ownership in many instances.

Land use posed the most difficulty for researchers, largely because of the inadequacy of information on county tax rolls. Land use was deemed an integral part of the landownership study because of its usual relationship with ownership. It is an indicator of the value and purposes of landownership as well as of likely future development in the local community. The following land use categories were developed to reflect the variety of land use designations thought to be recorded in county tax books.

1. commercial/industrial referred to land designated as commercial and/or industrial for taxation purposes.
2. agricultural plain referred to land designated for purposes of taxation as agricultural, where the land was used for pasture or other uncultivated purposes.
3. agricultural prime denoted land designated for purposes of taxation as agricultural, where the land is used for cultivated crops.
4. woodland/forest signified land designated as woodland, timber, or forests for taxation purposes.
5. residential referred to property that was listed for tax purposes as residential (on which the owner maintains a permanent or part-time residence).
6. recreational described property whose use was designated as for some recreational purpose (e.g., park, wilderness area).
7. mineral under development referred to land whose use was designated for purposes of taxation as mineral and whose minerals were in the process of being mined.
8. minerals not under development denoted land whose use was designated as mineral for purposes of taxation and whose minerals were not currently in the process of being mined.

Field researchers were instructed to code for primary use as well as any secondary uses so as to account for land with multiple uses.

While the land use categories were developed to include most types of land use expected in rural Appalachia, the general lack of land use information in county tax books diminished their value. Generally speaking, land use designations were recorded only partially, if at all. Both the quality of the data and the adequacy of our categories varied from county to county and state to state. At times additional land use information could be obtained from various public agencies, but it was not usually parcel specific. In sum, while the data on land use was very informative, it did not turn out to be sufficient for any sophisticated analysis of the relationship between landownership and land use.

Total mineral acres referred to the actual acres designated as mineral regardless of the type of mineral rights owned. The extensive severance of minerals from surface land in Appalachia, particularly in the central Appalachian coal counties, made this category necessary. Also, previous studies and current projections for energy development in the region give particular importance to the ownership of mineral rights. It is difficult to be sure of current information on mineral ownership, since it is constantly changing, except in a few central Appalachian counties with long established patterns of concentrated coal ownership. Even there, frequent transactions between large energy conglomerates change the name and address of the owner, though seldom the type.

The Task Force had assumed that mineral acreages would be readily available in reasonably accurate form from the county tax rolls. This was not to be the case. In Kentucky, for instance, county tax rolls did not include acreage figures and assessments for minerals. A few counties did have the information on the rolls, but many did not. Often it was possible to obtain limited information on coal and other mineral ownership, but the actual ownership of particular tracts was difficult to determine. To complete this information, it was necessary to copy the data from computer printout sheets issued to each county by the State Department of Revenue. It was still not initially possible to obtain these figures in five of the survey counties. As a result, total mineral acreages were under-recorded in our aggregate ownership analysis in Kentucky. Mineral acreages were later obtained for some of these counties and included in the study's county profiles, though not in the aggregate summaries of ownership.

In Alabama, severed mineral rights also often go unrecorded in the county tax books. When they were recorded, they were usually not even designated as mineral rights. However, such ownership could be identified by its low assessment, since the mineral assessment was so far below that on surface land. In the other states where mineral ownership was present, it was usually recorded in the tax books. Its accuracy was always open to question, though, since those mineral acreage figures often depended upon the willingness of owners to disclose their holdings.

Mineral type was included to take into account the variations in mineral ownership in Appalachia. The categories designated were coal, gas and oil, other, and combination. The *other* category was to account for mineral ownership in noncoal sections of Appalachia where the minerals might include zinc, lead, mica, olivine, stone, and so on. Two problems arose in the determination of mineral type. First, most counties do not record types of mineral, but refer simply to mineral ownership. Second, many of the mineral rights are leased rather than owned outright and were thus not recorded in the real property tax books.

Land, building, and mineral values categories were designed to ascertain the appraised values of land, buildings, and minerals owned by any particular owner. With this information, researchers could analyze the relationship between the value of a given holding and the actual taxes paid on that holding. Variations in recording systems and assessment practices of county tax assessors posed some problems here. The presence of appraised values in county tax books turned out to be less than universal.

The major problem in this regard was that some county records included only assessed values, while others listed the appraised value. In those counties in the study in which assessments were recorded, it was necessary to convert the assessed value to an appraised value. This was done by identifying an assessment-to-appraisal ratio for each of the counties. In most counties this ratio is relatively fixed and thus easy to determine. In others it is complicated by the flexibility given the local assessor in establishing assessments.

One other significant problem became evident while documenting the values of land, building, and mineral properties. In many counties these values were not recorded separately on the tax rolls (e.g., North Carolina counties). The result was an initial over-estimation of value per acre for land and an under-estimation of building values. A similar problem arose in some of the coal counties in which mineral and land values were not recorded separately (e.g., some West Virginia counties). While acres were summed separately as surface or mineral, the combined value was coded under surface. Thus, the total surface was overvalued and the total mineral undervalued, initially distorting value/acreage computations.

The tax paid category was designed to document the total taxes paid on any given property holding, whether land, buildings, or mineral. In actuality the figure recorded was the amount of tax charged, since the researchers could not determine whether owners had remitted any or all of the amount. Caution was necessary to avoid the inclusion of taxes assessed on equipment and personal property. This was usually self-evident since researchers were only using the real property books. However, in a few instances it was difficult to tell with certainty whether the tax assessed included both real and personal property.

In many cases it was impossible to determine the taxes assessed on utility property since it did not appear on the county tax rolls. This was particularly true where the relevant state agencies deemed such information confidential. In West Virginia, for instance, there were only partial utility acreage figures along with their valuation. These figures were included in our corporate totals, but the taxes were not. Both acreages and taxes assessed on utility owned property were missing from several Alabama counties.

The name, address, and zip code of the owner was recorded directly from the tax rolls. These categories were essential for determining the type and residence of the owners as discussed earlier in this section. The inclusion of zip code identification was intended to make possible the grouping of owners according to zip code, thus establishing residence patterns for nonlocal owners. While this was not included in our analysis, it could be done relatively easily by other researchers.

In conclusion, the coding categories used in the survey phase of the research met both the needs of local citizen researchers and the demands of computer analysis. Where data was missing, researchers had to be innovative in the use of other sources that would supplement the data found on county tax rolls. This was necessary to obtain information that should have been on the tax rolls. The experience of having to look elsewhere for data that should have been there convinced the Task Force of the general need for better landownership and taxation records in the counties studied. In spite of these limitations, the study represents an impressive compilation of land and mineral ownership in those counties.

Whereas the survey phase of the study documented landownership, use, and taxation patterns, the case study phase explored the impacts that those patterns have on local communities. Case study counties were chosen for the same reasons as the survey counties: that they represented various landownership patterns and that there was sufficient citizen interest to complete the studies and use their findings. In retrospect, the use of the findings by citizens varied considerably from county to county. The state task forces chose the case study counties from the survey counties in their respective states. Using the above criteria, nineteen counties were selected for investigation of the impacts of landownership patterns, current and past trends, and local response to those patterns and trends.

The case studies were designed to be open-ended and exploratory investigations of the relationship between landownership patterns and other variables at the county level. Selection of those variables was informed by the findings of previous studies, other literature about Appalachia, and, most importantly, the experiences of local citizens in those counties. Among the variables considered were housing, education, economic development, land use, social services delivery, local politics, and so on.

Information for the case studies came from three basic sources: the landownership survey, available documents, and interviews with county residents. Interviews were considered most critical to the elaboration of relationships suggested by the survey and available data as well as to the identification of other relationships not readily apparent in those sources. Interviewees were chosen on the basis of two general, nonexclusive criteria: (a) that they occupy a position in the county or have experience that would indicate familiarity with the dominant landownership patterns and related issues; (b) that as a group they be representative of the different segments of the local population (e.g., property owners and nonowners, business and labor, etc.). In neither instance were interviewees chosen at random. Thus, they did not represent the local population in other than the ways described above. Final selection of interviewees was left to the field researcher in the case study county after training sessions at the regional and state levels.

The precise questions asked interviewees varied from county to county, depending upon the dominant landownership patterns and land-related issues in any given county. However, the impact of landownership on the following factors was to be investigated in all the counties: economic development, particularly diversification and services infrastructure; community development, especially housing, environmental quality, and social services; and fiscal development, including sources of county revenue, tax rates, and county budget allocations. Case study researchers were also to investigate past and current trends in land tenure patterns in the county. The interviewers were further encouraged to use an open-ended format, conducive to the exploration of additional and unexpected relationships.

The major problem with the interview technique as a mode of investigation came from the controversial nature of landownership and the control of natural resources wealth in Appalachia. In some areas interviewees were reluctant to discuss the specifics of landownership and its impacts on their county. This was particularly true in counties where the owners of the land and natural resources also controlled the local economy, including a large proportion of local job opportunities. In spite of such limitations, there were numerous outstanding examples of open and frank discussion of such issues in most states.

A second source of information for developing county case studies was that myriad of available documents containing aggregate data about the county. Training sessions helped field researchers identify these documents and where they were likely to be found. Sources used included publications covering the history and development of the county, census documents, regional planning documents, county development plans, county budgets, local and regional newspapers, agricultural extension publications, and so on. Often this information was not as current as would have been desired, but it nevertheless became a useful resource for analyzing the impacts of landownership on the local level.

The survey phase of the ownership study provided the third source of information for the case studies. It served to orient the case study investigation in the sense that the survey identified dominant patterns of ownership and assessment of land in the county. In other words, the researcher knew a great deal more about what to investigate, what questions to ask, and whom to interview. Also, the survey data provided a means by which to verify information obtained in some of the interviews.

Over 100 variables were compiled and coded for each of the eighty counties. This data was grouped into four basic categories: land use, economic impacts, community impacts, and fiscal impacts (see Table A-4). Much of this information was provided by the data bank of the Appalachian Regional Commission, while other information came from such available sources as the *Agricultural Census, Census of Governments, Housing Census,* and so on. The data was collected for county units so as to facilitate the correlations between them and the relevant landownership findings. While its quality varied considerably and some of it was dated, it nevertheless provided us a means of correlating landownership patterns with various

Appendix Two 171

indicators of community and individual well-being. When this analysis was combined with the findings of the case studies, the complex picture of the impacts of landownership became much clearer than if either had been used alone.

The sixteen variables on the coding sheet were collected for 55,000 parcels of property, resulting in over 800,000 pieces of information. This data was keypunched and computerized at the computer center of Appalachian State University. Printouts of the information were then provided for each county and returned to the state coordinators for verification. Where necessary, further research was done on the ownership and residence of major owners. Corrected data was then keypunched again and entered into the computer for analysis.

The analysis of the data included: computing the number of owners and percentage of sample and county owned for surface and mineral owners by nature of owner (individual, corporate, and public) and by residence of the owner (in-county, out-of-county, out-of-state); ranking the owners according to size; computing indices of concentration (i.e., the distribution of acreage among the owners); calculating the taxes paid per acre for surface, minerals, and buildings; analyzing the distribution of taxes paid among owners by nature, residence, and size; and sorting acres and owners by land uses and mineral types. In this process of aggregating and analyzing data, a number of new variables were created, in addition to those on the coding sheet.

1. The first major task in analyzing the data involved the fact that one owner could own a number of parcels, in one county or across counties and even states. Since ownership rather than parcelization was the problem being studied, these parcels and the related data had to be collapsed or summed by owner. Often, though, the same owners would be referred to in different fashions (e.g., Consolidated Coal or Consol) and the computer would list them as separate owners. At the county level attempts were made in the corrections process to standardize names of the same owner. Then, an owner's listing was created, producing 33,000 owners from the original 55,000 parcels.

The collapsing process was more difficult when dealing with several counties or states. As it turned out, it was not possible to do this by computer at the time the information was needed. However, it was possible to compute manually the combined holdings of these owners and their subsidiaries in cases where they were listed as top owners in a given state. However, when aggregate calculations were done, involving such factors as number of owners or percent of owners, it was not always possible to combine holdings of the same owner, where the owner's name or title varied. As a result, these calculations *overstate* the number of owners and *understate* the degree of concentration.

2. Calculations of the percentage of land owned by various types of owners and for various types of use were done in two ways. The first involved the percentage of the acreage owned *in the sample*. The second involves the percentage of the total county surface, of the total surface of

the counties surveyed in the state, in each type, or in the whole sample. In the case of mineral rights, percentages are given as the percentage of minerals in the sample or as a percentage of the county land surface. This was necessary because the total number of mineral acres in any given county could not be determined. Overall, the land in the sample accounts for fifty-three percent of the land in the eighty counties, while the mineral acres are the equivalent of twenty-two percent of the surface of those counties. Care was taken throughout the study to specify which percentages were being used as indicators.

One should recognize, however, that where percentages of a county are given, these refer only to the acres in the sample as a percentage of the county. For instance, when the study reports that thirty percent of the land in a county is corporately owned, it refers only to the land in the sample (i.e., above the twenty acre limit) as a percentage of the county's total surface. This likely *understated* the amount of land which actually is corporately owned, since corporations own additional acreage too small to be included in our sample.

3. In the study, concentration—the degree to which land is held amongst few owners or dispersed among several owners—was measured in two ways. The simplest index was obtained by dividing the percentage of land owned by a certain top percent of owners in the sample by the percentage of land owned by a certain bottom percent of owners in the sample. The higher the index, the greater the concentration: the lower the index, the lower the concentration. Generally, this was measured as a ratio of the amount of land owned by the top twenty-five percent of owners in the sample divided by the amount of land owned by the bottom twenty-five percent of the owners in the sample.

The second and more technical index used was the Gini co-efficient, a standard measurement for the distribution of income based on the Lorenz curve. This method was applied to the measurement of concentration in landownership by Gene Wunderlich in 1958. In his terms, "The area between the Lorenz curve and the line of perfect equality represents the degree of concentration.... The Gini ratio of concentration is simply the ratio of the area between the Lorenz curve and the line of perfect equality to the total area of the triangle formed by the two axes and the line of perfect equality."[13]

As used, both indices *understate* the degree of concentration of ownership actually present. First, the concentration of ownership can be given only for the owners sampled, not for all owners in a county. Second, on the aggregate level, it was not always possible to combine all parcels owned by the same owner if that ownership occurred in several counties or states.

4. Property values recorded on the tax books were collected for surface, minerals, and buildings. In some cases, these values represented the full appraised value of the property; in other cases, they represented only an assessment, or a percentage of the appraised value. Attempts were made to standardize the values by multiplying the assessments by 1/assessment ratio. However, the assessment ratio was often arbitrarily determined by the

Appendix Two

assessor and the appraisal did not always reflect the full value of the property. The effort to standardize the meaning of values by using the assessment ratio given in the *Census of Governments* was only partially successful, since the ratio was not available for all counties. Because of these difficulties, discussion of taxes usually refers to taxes per acre as derived from the tax rolls, rather than on appraised values.

5. The inadequacy of the appraisal as a method of comparison meant that new calculations had to be done to obtain a tax per acre figure. The problem here arose from the fact that the "tax paid" column on the tax books reflects the taxes billed on the sum value of the surface, minerals, and buildings. In order to determine what proportion of the amount could be applied to which component, total valuation in each county was divided by total taxes collected in order to get a tax rate for the sample in the county. The tax rate is an "internal" rate for the sample and may not correspond precisely with the tax rate used for actual taxes. The rate was then applied to the value of each component—surface, mineral, buildings—to determine what proportion of the taxes paid came from each category. Those figures were then divided by the total number of surface acres, mineral acres, or building lots in order to determine the surface tax per acre, mineral tax per mineral acre, or building tax per building lot.

In the case of mineral taxes, this procedure means that the figure of mineral tax per mineral acre obviously refers only to the value of those mineral acres which are listed for tax purposes. Where minerals are not listed at all, as is often the case, or where their value is reflected in the surface value, as is sometimes the case in West Virginia, the mineral tax per mineral acre will *overestimate* the actual tax on the mineral rights.

6. In order to get a yardstick for measuring the proportion of property taxes paid by types of landowners, a measure of the total property taxes collected in the county was needed. Since this information was not collected during the field research, the total property taxes figures were taken from the 1977 *Census of Governments*. The percentage of property taxes paid by a given category of owners was determined by dividing the total taxes paid in the survey by the category of owners by the *Census of Government* figure for the total property taxes paid. The measure was crude for two reasons. First, while the survey data provided the total real estate taxes, the *Census of Governments* figure included *all* property taxes, some of it assessed on intangible property. As a result, the yardstick figure must be understood to mean the amount of real estate taxes paid, as a percentage of all property taxes paid, not just real estate taxes. As such, it probably *understates* the actual tax burden. The second problem, however, may *overstate* the tax burden. While the *Census of Governments* figure was for 1976-77, the survey data was for two or three years later. During this interval taxes paid could have increased. Despite these difficulties, spot checks revealed the figure to be close to accurate with the exception of West Virginia. In that state total property taxes paid changed greatly between 1976-77 and 1978-79 in counties where new mineral taxation procedures were applied. To correct this for the West Virginia state and county profile analysis, the 1978-79 real

estate taxes totals were substituted for the *Census of Governments* data in order to give a more accurate estimate of actual tax burden.

Once computed, the aggregate data was then analyzed on four different levels: by county, by state, by sample (region), and by county type. In each case, the data used represented the sum of the totals for the survey counties in that category, i.e., the total surface acres in West Virginia refers to the total surface acres in the survey counties. However, the definition of absentee as being out-of-county or out-of-state remained in reference to the county, regardless of the unit of analysis. So, the percentage of land absentee owned in a particular state still means the percentage owned by owners not residing in the county where the holdings are located, rather than the percentage not residing in the state where the holdings are.

Past studies of land ownership indicate that ownership patterns and their impacts may differ according to the types of economic activity for which the land is used. Three types of land-based economic resources were expected to be particularly significant in predicting ownership patterns and their impacts in Appalachia. These three are: the level of coal reserves and coal production, level of agricultural use, and level of rural tourism and recreation. In order to test these relationships further, counties were typed according to the importance of these land based economies.

This typing was done by using measures deemed independent of ownership or use characteristics. In the case of coal counties, the level of known coal reserves was used as the measure. Some consideration was given to using levels of coal production, but this was not thought to measure the speculative importance of holding coal land for future exploitation. In the case of agricultural counties, the value of agricultural sales recorded in the 1974 *Census of Agriculture* was used. The case of tourism/recreation was more difficult as standard indicators were not readily available. Finally, using the *Census of Selected Service Industries,* a measure was developed which combined the percentage of service receipts in hotels, motels, trailer parks, and campgrounds *plus* the percentage of service receipts from amusements and recreation industry. The dominance of these types of service industries was taken as an indicator of the prevalence of tourism and recreation in the county's economy. Counties within each type were further characterized as to whether they ranked high or low with respect to coal reserves, agricultural sales, or the service receipts. This allowed for additional comparison within types and between types. It should also be noted that these categories are not mutually exclusive, since a given county could conceivably rank high on more than one of the indicators. This was not usually the case, however. None of the above indicators of county type are without problems and thus should be seen only as one attempt to categorize counties for purposes of analyzing the impacts of landownership. (See Table A-1 for additional details of this typology).

County case studies were subjected to a thorough process of verification and review before their inclusion as part of the state reports and use in the regional analysis. Initial drafts of these case studies were written by the local researcher who had conducted the interviews and collected appropriate county level data. He/she was assisted in this process by the state coordinator and task force as well as by the regional staff. Once the initial draft was completed, it was forwarded to the state coordinator for review, editing, verification, and the addition of supplementary information. Then, the case study was forwarded to the regional staff for examination of its content, format, and comparability to other case study drafts. At this time any statistical data in the case studies was checked against that collected at the regional level and, where necessary, corrected. Perceptions provided by interviewees were also checked against other sources when possible, though they were treated as important reflections of local opinion even if they could not be verified. Once edited by the regional staff, the case study was returned to the state coordinator with corrections, questions, suggestions, and additional relevant information. The state coordinator was then responsible, in collaboration with the original author, for writing a revised draft of the study.

Case studies were utilized in two distinct ways in the overall context of the landownership study: as entities unto themselves and as essential sources of information for the state and regional reports. First, the case studies provided our only comprehensive accounts of the impacts of landownership on local communities. The combination of data sources used provided a unique opportunity to examine both the objective account of those impacts and the perceptions of local residents about them. Ultimately, it is at the local level where the impacts of ownership, whether positive or negative, are felt. The county case studies stand as documentation of how local residents respond to those impacts. Unfortunately, it was impossible to include the case studies in this volume. They were, however, reproduced in their entirety in the state reports.

Second, with respect to state and regional reports, the case studies helped define the pertinent issues relating to landownership while offering personal confirmation of their impacts on the daily lives of local citizens. Such insights were indispensable in the final decisions about impacts that deserved extended discussion in the state and regional reports. These issues were identified after a review of all case studies, with particular focus on the issues mentioned by local residents.

Once the broad impact areas were identified, the case studies were subjected to content analysis in such a way that all interview materials dealing with a particular impact area were elicited from them. Information on landownership and taxation was also excerpted. Once these were compiled, they were incorporated into the relevant arguments in the state and regional discussions of land ownership and its impacts. So used, case study findings became an integral part of the overall analysis.

In this phase of the analysis correlations between certain landownership characteristics and socioeconomic variables were examined for all eighty counties in the study. Landownership characteristics were based upon the percentage of the county owned by different types of owners rather than the percentage of the sample so owned. Ten landownership characteristics were developed as independent variables to correlated with the socioeconomic or dependent variables, the aim being to discover possible impacts of absentee, corporate, and government holdings. These ownership variables were:

1. Percentage of county in corporate ownership.
2. Percentage of county in public (government and private non-profit) ownership.
3. Percentage of county in corporate and public ownership.
4. Percentage of county owned by out-of-county, but in-state owners.
5. Percentage of county owned by out-of-state owners.
6. Percentage of county owned by non-local owners (4 + 5).
7. Percentage of county owned by absentee corporate and government owners.
8. Percentage of county owned by absentee + corporate + government owners (i.e., all owners coded but local individuals).
9. Percentage of county owned by nonlocal individuals.
10. a. Concentration of ownership (percent of land owned by the top 25 percent divided by percent owned by the bottom 25 percent).
    b. Concentration of ownership (Gini co-efficient).

To test the impact of mineral ownership, the same variables were developed for it as a percentage of surface ownership. Where these are used, it should be remembered that mineral ownership records are incomplete.

It was expected that in some cases the ownership of surface and mineral *combined* would strengthen the effect that either would have alone. This was primarily anticipated where mineral rights are severed from surface rights. Corporate or absentee control can involve both surface and mineral ownership. To examine this possible combined effect, the above ownership indices were also developed for percentage of surface owned *plus* percentage of mineral owned. As in the case of mineral rights, this "index of resource control" must be used with the recognition that mineral rights data were limited.

When correlating ownership characteristics with the aggregate data, emphasis was placed on correlations with land use, economic, social, and fiscal (county finance) indicators. Three correlation measures were initially used: Pearson's R, Spearman's, and Kendall's Tau. Of these, only Pearson's R, being the most stringent test for determining the relationship between two variables, was used in the analysis presented in the study.

In determining the significance of relationships, the following general criteria were used.

1. The Pearson's R correlation was deemed significant only if (a) the level of probability that the correlation was not random was less than .05 (in most cases, it was less than .01) and (b), in general, if the strength of the relationship was greater than .30. As a rule of thumb, relationships in the .300-450 range were considered significant but weak; in the .451-600 range, strong; and over .600 to be very strong.

2. Isolated correlations, even if significant by the above criteria, were not used to draw conclusions. Due to the number of dependent variables used, relationships were only considered significant if a pattern could be found among the various variables. For example, a relationship that was consistent across all types of absentee holdings, singly or in combination, was given more weight than those occuring with only one type of absentee ownership.

Even with these restrictions, significance of relationships must still be qualified because of limitations in both the aggregate data and the courthouse records. Nevertheless, these correlations do point out important impacts of land ownership and when corroborated by case study and other data add substantially to our knowledge of such impacts in the Appalachian region.

APPENDIX THREE

Annotated Bibliography

Steve Fisher

Special thanks to Louise Fachilla, John Gaventa, Virginia Groseclose, Mary Harnish, and the Highlander Center for help in preparing this bibliography. A preliminary version was published in May 1979 by the Highlander Research and Education Center, New Market, Tennessee. The version used here was revised in July 1980.

CONTENTS

1. General Works   179
   History   179
   Landownership   182
   Property Taxation   186

2. Coal Landownership and Property Taxation   187
   Appalachian Regional and State Studies   187
   Impact Studies   190

3. Recreation Landownership and Second-home Development   194

4. Farm Landownership   198
   Farm Landownership and Taxation   198
   Family Farm Issues   201

5. Government Landownership, Especially National Forests   206

6. Federal and Private Dam Builders   211

7. Land-reform Proposals and Strategies   215

Appendix Three

ABBREVIATIONS

AER	Agricultural Economic Report
AIB	Agricultural Information Bulletin
ARC	Appalachian Regional Commission
ERS	Economic Research Service [of the USDA]
ESCS	Economics, Statistics, and Cooperatives Service [of the USDA]
F	Steve Fisher, ed. *A Landless People in a Rural Region: A Reader on Land Ownership and Property Taxation in Appalachia.* New Market, Tenn.: Highlander Center, 1979.
IAAO	International Association of Assessing Officers
JHP	Johns Hopkins University Press
LJ&A	Helen Lewis, Linda Johnson, and Don Askins, eds. *Colonialism in Modern America: The Appalachian Case.* Boone, N.C.: Appalachian Consortium Press, 1978.
MLW	*Mountain Life and Work*
MP	Miscellaneous Publication
RFF	Resources for the Future
TRED	Committee on Taxation, Resources, and Economic Development
TVA	Tennessee Valley Authority
Univ.	University
USDA	United States Department of Agriculture
USFS	United States Forest Service
USGPO	United States Government Printing Office
VPI	Virginia Polytechnic Institute and State University
WVU	West Virginia University

1. GENERAL WORKS

History. This section provides only a sampling of the major works related to the historical development of landownership patterns in the United States. Several key works on the history of the conservation movement are included, but conservation and environmental texts are omitted. Regional and state histories unrelated to Appalachia are not listed.

Abernethy, Thomas P. *Western Lands and the American Revolution.* N.Y.: Appleton, 1937; repr. N.Y.: Russell & Russell, 1959. Political effects of the trans-Appalachian westward movement on the land policies of the British and American governments.

Abrams, Charles. *Revolution in Land.* N.Y.: Harper, 1939. Interesting historical critique of land tenure in the United States as a battle between industry and agriculture. Concludes that industry won and calls for a program of land nationalization.

Carstensen, Vernon, ed. *The Public Lands: Studies in the History of the Public Domain.* Madison: Univ. of Wisconsin Press, 1963. Anthology of historical articles.

Chandler, Alfred N. *Land Title Origins: A Tale of Force and Fraud.* N.Y.: Robert Schalkenbach Foundation, 1945. State-by-state account of how "unscrupulous men of great political power and influence" initially gained control of the public domain.

Clawson, Marion. *The Bureau of Land Management.* N.Y.: Praeger, 1971. Historical treatment of the major resource bureau in the Dept. of Interior.

Dick, Everett. *The Lure of the Land: A Social History of the Public Lands from the Articles of Confederation to the New Deal.* Lincoln: Univ. of Nebraska Press, 1970. Examines the human side of the federal process of land distribution.

Ellis, David M., ed. *The Frontier in American Development: Essays in Honor of Paul Wallace Gates.* Ithaca: Cornell Univ. Press, 1969. Essays on the history of public land and land disposal.

Gates, Paul W. *History of Public Land Law Development.* Written for the Public Land Law Review Commission. Washington., D.C.: USGPO, 1968. Examines the historical development of present and past public land laws.

―――. "Research in the History of American Land Tenure: A Review Article." *Agricultural History* 28 (July 1954): 121-26.

―――, ed. *Public Land Policies: Management and Disposal.* N.Y.: Arno, 1979. Focuses on the major problems of land administration throughout American history.

Harris, Marshall. *Origin of the Land Tenure System in the United States.* Ames: Iowa State College Press, 1953. Focuses on the tenure process during the two centuries of the Colonial era.

Hays, Samuel P. *Conservation and the Gospel of Efficiency: The Progressive Conservation Movement, 1890-1920.* Harvard Historical Monograph, No. 40. Cambridge: Harvard Univ. Press, 1959. Standard work, which argues that conservation did not develop as a mass movement but as a scientific movement led by specialists loyal to professional ideals.

Hibbard, Benjamin H. *A History of the Public Land Policies.* N.Y.: Macmillan, 1924; repr. Madison: Univ. of Wisconsin Press, 1965. Reference work sketching the historical development and disposition of the public domain.

Ise, John. *Our National Park Policy: A Critical History.* Pub. for RFF. Baltimore: JHP, 1961; repr. N.Y.: Arno Press, 1979. Contains history of each park and of each National Park Service administration since 1916.

Josephson, Matthew. *The Robber Barons: The Great American Capitalists, 1861-1901.* N.Y.: Harcourt, 1934. Includes portraits of "barons" who made their fortunes through land control and speculation.

Livermore, Shaw. *Early American Land Companies: Their Influence on Corporate Development.* N.Y.: Commonwealth Fund, 1939. Focus on pre- and post-revolutionary land companies and their activities.

Moyer, D. David; Harris, Marshall; and Harmon, Marie B. *Land Tenure in the United States: Development and Status.* AIB-338. Washington, D.C.: ERS, USDA, June 1969. Examines historical origins and trends in landownership and control.

Nixon, Edgar B., ed. *Franklin D. Roosevelt and Conservation, 1911-1945.* 2 vols. Hyde Park, N.Y.: General Services Administration, 1957; repr. N.Y.: Arno,

1972. Indexed, annotated collection of the most important presidential correspondence concerning conservation. Valuable source of conservation history during the New Deal.

Peffer, E. Louise. *The Closing of the Public Domain: Disposal and Reservation Policies, 1900-50.* Stanford: Stanford Univ. Press, 1951. Relates steps by which the concept of public domain has veered from one of land held in escrow pending transfer of title, toward one of reservations held in perpetuity in the interest of the collective owners, the American people

Penick, James L., Jr. *Progressive Politics and Conservation: The Ballinger-Pinchot Affair.* Chicago: Univ. of Chicago Press, 1968. Explores 1910 political controversy that split the conservation movement.

Petulla, Joseph M. *American Environmental History: The Exploitation and Conservation of Natural Resources.* San Francisco: Boyd & Fraser, 1977. History of anticonservationist policy and practice, and a study of the development of a political economy whose chief imperative is growth.

Puter, S. A. D., in collaboration with Horace Stevens. *Looters of the Public Domain, Embracing a Complete Exposure of the Fraudulent Systems of Acquiring Titles to the Public Lands of the United States.* Portland, Or.: n. p., 1908; repr. N.Y.: Arno, 1972. Discloses the techniques used by individuals to defraud the government of public lands.

Richardson, Elmo R. *Dams, Parks and Politics: Resource Development and Preservation in the Truman-Eisenhower Era.* Lexington: Univ. Press of Kentucky, 1973. Analysis of the conservation issues of the 1950s.

―――. *The Politics of Conservation: Crusades and Controversies, 1897-1913.* Univ. of California Publications in History, Vol. 70. Berkeley: Univ. of California Press, 1962. On the conservation movement and its role in Western and national politics, culminating in the Ballinger-Pinchot controversy.

Robbins, Roy M. *Our Landed Heritage: The Public Domain, 1776-1936.* Princeton: Princeton Univ. Press, 1942; repr. Lincoln: Univ. of Nebraska Press, 1962. Standard history of federal land policy.

Rohrbough, Malcolm J. *The Land Office Business: The Settlement and Administration of American Public Lands, 1789-1837.* N.Y.: Oxford Univ. Press, 1968. General history of federal land policy and its administration by the General Land Office.

Sakolski, Aaron M. *The Great American Land Bubble: The Amazing Story of Land-Grabbing, Speculations and Booms from Colonial Days to the Present Time.* N.Y.: Harper, 1932. Readable account of how national landownership patterns developed.

―――. *Land Tenure and Land Taxation in America.* N.Y.: Robert Schalkenbach Foundation, 1957. Traces the evolution of our present land-tenure system and evaluates the probable effect of various land taxation proposals upon land use and social development.

Smith, Frank E. *The Politics of Conservation.* N.Y.: Pantheon, 1966.

Smith, Frank E.; Foss, Phillip O.; Doherty, William T., Jr.; and Divorsky, Leonard B., eds. *Conservation in the United States: A Documentary History.* 5 vols. N.Y.: Chelsea House, 1971. Excerpts of legislation, reports, and speeches related to conservation issues.

Swain, Donald C. *Federal Conservation Policy, 1921-1933.* Univ. of California Publications in History, Vol. 76. Berkeley: Univ. of California Press, 1963. Sets the stage for the study of conservation policy during the New Deal years. A thorough study that covers national forests, national parks, reclamation, minerals, and water power.

Treat, Payson Jackson. *The National Land System, 1785-1820.* N.Y.: E. B. Treat Co., 1910; repr. N.Y.: Russell & Russell, 1967. Detailed analysis of the origins of the land system and the twenty-year credit period in the sale of the public domain.

Warne, William E. *The Bureau of Reclamation.* N.Y.: Praeger, 1973. History of this agency and its projects.

Watkins, T. H., and Watson, Charles S., Jr. *The Land No One Knows: America and the Public Domain.* San Francisco: Sierra Club Books, 1975. Traces the progressive loss of our common land inheritance to private landholders.

Landownership. This section identifies the more significant texts, collections of readings, general surveys, and bibliographies concerned with landownership in Appalachia and the United States. Several works focusing on special topics related to landownership are also included. Works focusing on black and foreign ownership patterns are listed in Section 4. No effort is made to list works concerned primarily with conservation, natural resources, or land use issues.

Andrews, Richard N. L., ed. *Land in America: Commodity or Natural Resource?* Lexington, Mass: Lexington Books, 1979. Essays on the history of land in America, people's perceptions and images of land ownership, and government and the land. Includes a useful bibliography.

Barlowe, Raleigh. *Land Resource Economics: The Economics of Real Estate.* 3rd ed. Englewood Cliffs, N.J.: Prentice-Hall, 1978. A leading textbook in the field.

Barnes, Peter, ed. *The People's Land: A Reader on Land Reform in the United States.* Emmaus, Pa.: Rodale, 1975. Excellent collection of articles, studies, and statements dealing with the issues of landownership and land use.

Barnes, Peter, and Casalino, Larry. *Who Owns the Land? A Primer on Land Reform in the United States.* Berkeley, Cal.: Center for Rural Studies, 1972. Summarizes key land issues and suggests appropriate remedies.

Behrens, John O.; Moyer, D. David; and Wunderlich, Gene. *Land Title Recording in the United States: A Statistical Summary.* State and Local Government Special Studies, No. 67. Washington D.C.: USDA and U. S. Dept. of Commerce, March 1974. Survey of real estate transfer procedures.

Bertrand, Alvin L., and Corty, Floyd L., eds. *Rural Land Tenure in the United States: A Socio-Economic Approach to Problems, Programs, and Trends.* Baton Rouge: Louisiana State Univ. Press, 1962. Uses an interdisciplinary approach to present current research.

Bingham, Edgar. "Appalachia: Underdeveloped, Overdeveloped, or Wrongly Developed?" *The Virginia Geographer* 7 (Fall-Winter 1972): 9-12.

Bosselman, Fred; Callies, David; and Banta, John. *The Taking Issue: A Study of the Constitutional Limits of Governmental Authority to Regulate the Use of*

Appendix Three 183

Privately-Owned Land Without Paying Compensation to the Owners. Washington, D.C.: Council on Environmental Quality, 1973. Examines the political and legal history of our constitutional powers affecting land.

Boxley, Robert F. *Landownership Issues in Rural America.* ERS-655. Washington, D.C.: ERS, USDA, April 1977.

Branscome, Jim. "If Appalachia Is to Survive, Land Reform Is a Must." *Mountain Eagle* (4 January 1973). Also in *MLW* (May 1973): 11-14; *Peoples Appalachia* 3 (Spring 1973): 32-33; and F, pp. 24-25. Identifies the major land issues in Appalachia and discusses several reform proposals.

Browning, Frank. *The Vanishing Land: The Corporate Theft of America.* N.Y.: Harper, 1975. Shows how each year more and more land is given over to major banks, manufacturers, and insurance companies.

Burke, Barlowe, Jr., and Wunderlich, Gene, eds. *Secrecy and Disclosure of Wealth in Land.* Washington, D.C.: Farm Foundation in cooperation with USDA, 1978. Examines some of the major ethical, legal, and economic issues of securing information about who owns the land.

Carruth, Eleanore. "Look Who's Rushing into Real Estate." *Fortune* (October 1968): 160-63+. Discusses how ITT, Westinghouse, and other large corporations are investing in land.

Center for Rural Affairs. *Land Tenure Research Guide.* Walthill, Nebr., n.d. Guide to help people ask the right questions of public employees who manage the offices where information on land tenure is kept.

Chasen, Daniel Jack. *Up For Grabs: Inquiries into Who Wants What.* Seattle: Madrona, 1977. Examines some of the effects of the private ownership of land and water.

Clark, Mike. "How Can You Buy or Sell the Sky?" *Mountain Eagle* (23 June 1977). Discusses the importance of land to Appalachia's future.

Clawson, Marion. *America's Land and Its Uses.* Pub. for RFF. Baltimore: JHP, 1972. Survey of major facts and issues related to land and land policy.

Clawson, Marion; Held, R. Burnell; and Stoddard, Charles H. *Land for the Future.* Pub. for RFF. Baltimore: JHP, 1960. Considers the changing uses of land in America in the past, at present, and in light of expectations extending to the year 2000.

"The Corporate Land Rush of 1970." *Business Week* (29 August 1970): 72-77. Describes how more and more big companies are moving into real-estate development.

Denman, D. R.; Switzer, J. F. Q.; and Sawyer, O. H. M. *Bibliography of Rural Land Economy and Land Ownership, 1900-1957: A Full List of Works Relating to the British Isles and Selected Works, from the United States and Western Europe.* Cambridge: Dept. of Estate Management, Cambridge Univ., 1958. A useful bibliography.

Ely, Richard T., and Wehrwein, George S. *Land Economics.* N.Y.: Macmillan, 1940; repr. Madison: Univ. of Wisconsin Press, 1964. One of the most influential books in the history of land economics.

Fellmeth, Robert C. *Politics of Land: Ralph Nader's Study Group Report on Land Use in California.* N.Y.: Grossman, 1973. Found that twenty-five landowners hold more than 61 percent of California's private land.

Finger, Bill; Fowler, Cary; and Hughes, Chip. "Special Report on Food, Fuel, and Fiber." *Southern Exposure* 2 (Fall 1974): 145-210. Valuable statistical summaries.

Fisher, Steve. "Appalachians as 'Redskins': The Assault on the Land Continues." *Mountain Review* 4 (April 1979): 4-6. Examines the extent and significance of the assault on the land in Appalachia.

———, ed. *A Landless People in a Rural Region: A Reader on Land Ownership and Property Taxation in Appalachia.* New Market, Tenn.: Highlander Center, 1979. Includes excerpts from existing landownership studies along with articles that examine the impact of landownership patterns on the quality of life in Appalachia.

Frey, H. Thomas. *Major Uses of Land in the United States: 1974.* AER-440. Washington, D.C.: ESCS, USDA, November 1979. Summary of the extent and distribution of land used for crops, pasture and range, forestry, and various special purposes.

Friedenberg, Daniel M. "America's Land Boom: 1968." *Harper's* (May 1968): 25-32. Discusses how preferential tax treatment of speculators continues to create serious imbalances in land development.

Fugere, Joseph. "Corporate Invasion in Land Ownership." Washington, D.C.: Rural America, 1977. Broad overview of major corporate landowners.

Gilbert, Jess C., and Harris, Craig K. "Corporate Land Ownership and Rural Poverty: A Center-Periphery Model Applied to the Upper Peninsula of Michigan." Unpublished paper presented at the annual meeting of the Rural Sociological Society, San Francisco, 1-3 September 1978. Examines the relationship between landownership and rural poverty as a model of uneven spatial development.

Johnson, V. Webster, and Barlowe, Raleigh. *Land Problems and Policies.* N.Y.: McGraw-Hill, 1954. Text concerned primarily with land problems and policies rather than with the theoretical framework of land economics. Includes a useful historical perspective of American land policies and sections on agricultural requirements, land tenure, and land reform.

Jones, Lindsay, ed. *Citizen Participation in Rural Land Use Planning in the Tennessee Valley.* Nashville, Tenn: Agricultural Marketing Project, 1979. Good collection of essays that considers the problems Appalachians have in attaining access to land planning.

"Land." *Community Economics.* An Occasional Bulletin of the Center for Community Economic Development. Cambridge, Mass. May 1972. Special issue on landownership and abuse.

"Last Stand to Save the Land." *Peoples Appalachia* 2 (September-October 1971). Dated but important survey of Appalachian land issues.

Lewis, Douglas G. *Corporate Landholdings: An Inquiry into a Data Source.* ESCS Staff Report NRED 80-5. Washington, D. C.: ESCS, USDA, March 1980. Examines Securities and Exchange Commission reporting requirements for publicly traded corporations, aggregates the available data, assesses the data source, and suggests means to improve the data.

Lewis, James A. *Landownership in the United States, 1978.* AIB-435. Washington,

Appendix Three

D.C.: ESCS, USDA (April 1980). The data portray a broad picture of landownership characteristics at the national level and show some comparisons of landownership among regions.

Meyer, Peter. "Land Rush: A Survey of America's Land." *Harper's* (January 1979): 45-60 Also in F, pp. 4-19. Informative and up-to-date survey of who owns and controls the land.

Moyer, D. David. *Land Information Systems: An Annotated Bibliography.* Washington, D.C.: ESCS, USDA, 1978.

Moyer, D. David, and Dougherty, Arthur B. *Landownership in the Northeast United States: A Sourcebook.* Washington, D.C.: ERS, USDA, 1976. Reviews and evaluates the three major sources of landownership data for thirteen Northeastern states.

"The New American Land Rush." *Time* (1 October 1973): 80f. Examines the dimensions, causes, and consequences of the "new land rush" and focuses on some of the powerful individuals who determine how America uses its land.

Osborn, William C. *Paper Plantation: Ralph Nader's Study Group Report on the Pulp and Paper Industry in Maine.* N.Y.: Grossman, 1974. Reports that seven absentee corporations own 32 percent of Maine's 20 million acres.

Ottoson, Howard W., ed. *Land Use Policy and Problems in the United States.* Lincoln: Univ. of Nebraska Press, 1963.

"Our Promised Land." *Southern Exposure* 2 (Fall 1974). Excellent collection of essays on land issues in the South and Appalachia.

The Plow (November 1976). Special issue on land.

"Save the Land and People." *MLW* (May 1973). Provides a survey of groups working on land issues.

"The Shrinking Supply of Private Land." *U.S. News & World Report* (20 February 1978): 64-65. Discusses how state and local governments are buying up land at the rate of a million acres a year.

Smith, Charles L. *A Bibliography on Land Reform in Rural America.* San Francisco: Center for Rural Studies, 1974. Over a thousand references.

Stone, Christopher D. *Should Trees Have Standing? Toward Legal Rights for Natural Objects.* Los Altos, Cal.: William Kaufmann, 1974. Holistic argument that land and other natural objects, as well as persons, should have certain legal rights.

Timmons, John F., and Murray, William G., eds. *Land Problems and Policies.* Ames: Iowa State College Press, 1950; repr. N.Y.: Arno, 1972. Sixteen essays concerned with land problems and uses.

U.S., Congress, Senate. Subcommittee on Migratory Labor of the Committee on Labor and Public Welfare. *Farmworkers in Rural America, 1971-1972. Part 2: Who Owns the Land?* Hearings, 92nd Congress, 1st and 2nd sessions. Washington, D.C.: USGPO, 1972. Useful collection of statements and position papers on who owns the land.

USDA, ERS. *Our Land and Water Resources: Current and Prospective Supplies and Uses.* MP-1290. Washington, D.C.: USGPO, May 1974. American land and water resources are analyzed as a basis for projecting agricultural cropland and other land needs to the year 2000. Based on 1969 Census data. Includes a useful section on ownership and land use (pp. 20–32).

Wunderlich, Gene. *Facts About U.S. Landownership.* AIB-422. Washington, D.C.: ESCS, USDA, November 1978. Summarizes ownership issues and statistics. Discusses obstacles to getting details about landownership.

Young, John A., and Newton, Jan M. *Capitalism and Human Obsolescence: Corporate Control versus Individual Survival in Rural America.* Montclair, N.J.: Allanheld, Osmun, 1980. West Coast focus, with good chapters on the timber industry, the mining industry, and "the farm problem."

C. Property Taxation. These works deal with property taxation in general. Works concerned with property-tax issues as related to agricultural and forestlands are included in Sections 4 and 5.

Aaron, Henry J. *Who Pays the Property Tax? A New View.* Washington, D.C.: Brookings Institution, 1975. Useful overview of arguments for and against the property tax. Offers reform suggestions.

Advisory Committee on Intergovernmental Relations. *The Property Tax in a Changing Environment: Selected State Studies—An Information Report.* M-83. Washington, D.C.: March 1974.

Becker, Arthur P., ed. *Land and Building Taxes: Their Effect on Economic Development.* Proceedings of a symposium by TRED at the Univ. of Wisconsin, 1966. Madison: Univ. of Wisconsin Press, 1969.

Bernard, Michael M. *Constitutions, Taxation, and Land Policy.* Lexington, Mass: Lexington Books, 1979. Examines the key provisions in all state constitutions and the United States constitution that pertain to the limits of taxation on land.

Brandon, Robert M.; Rowe, Jonathan; and Stanton, Thomas H. *Tax Politics: How They Make You Pay and What You Can Do About It.* N.Y.: Pantheon, 1976. Excellent and easy-to-read coverage of the various dimensions of property taxation, along with a chapter on investigating property taxes. Includes a useful bibliographical essay.

Colby, Donald S., and Brooks, David B. *Mineral Resource Valuation for Public Policy.* Washington, D.C.: U.S. Dept. of Interior, Bureau of Mines, 1969. Manual for making the type of mineral resource valuation commonly required for such public policy problems as mineral leasing.

Gaffney, Mason, ed. *Extractive Resources and Taxation.* Proceedings of a Symposium Sponsored by TRED at the Univ. of Wisconsin, Milwaukee, 1964. Madison: Univ. of Wisconsin, 1967. Three major sections—theoretical foundations, economic institutions, and policy.

Greever, Barry. *Property Taxes: What to Look For and Where to Find it.* Mineral Bluff, Ga.: Cut Cane Assn., 1973. Useful guide to researching property taxes.

Holland, Daniel M., ed. *The Assessment of Land Value.* Proceedings of a symposium sponsored by TRED at the Univ. of Wisconsin, Milwaukee, 1969. Madison: Univ. of Wisconsin Press, 1970. Ten essays focusing on issues of site-value taxation and the problems of assessing land.

IAAO, ed. *Analyzing Assessment Equity.* Proceedings of a symposium conducted by the IAAO Research and Technical Services Department in cooperation with the Lincoln Institute of Land Policy. Chicago, 1977. Techniques for measuring and improving the quality of property-tax administration.

Jensen, Jens P. *Property Taxation in the United States.* Chicago: Univ. of Chicago Press, 1931. A classic work, which includes a history of property taxation.

Keene, John. *Untaxing Open Space.* Prepared for the Council on Environmental Quality. Washington, D.C.: USGPO, 1976. Includes a very useful bibliography on property taxation.

Keith, John. *Property Tax Assessment Practices: A Reference Book for the Assessor, Appraiser, Accountant, Attorney and the Student.* Monterey Park, Cal.: Highland, 1966. Closest thing to an assessor's desk manual.

Lindholm, Richard W. "Twenty-one Land Value Taxation Questions and Answers." *American Journal of Economics and Sociology* 31 (April 1972): 153-61. Provides a helpful explanation of site-value taxation.

———, ed. *Property Taxation, USA.* Proceedings of a symposium sponsored by TRED at the Univ. of Wisconsin, Milwaukee, 1965. Madison: Univ. of Wisconsin Press, 1967. Focus on business and industry and on special problems.

Lynn, Arthur D., Jr., ed. *The Property Tax and Its Administration.* Proceedings of a symposium sponsored by TRED at the Univ. of Wisconsin, Milwaukee, 1967. Madison: Univ. of Wisconsin Press, 1969. Focus on administrative organization assessment procedures and reforms.

———, ed. *Property Taxation, Land Use, and Public Policy.* Proceedings of a symposium sponsored by TRED at the Univ. of Wisconsin, Madison, 1973. Madison: Univ. of Wisconsin Press, 1976. Provides a current appraisal of, and offers alternatives to, the property tax.

Netzer, Dick. *The Economics of the Property Tax.* Washington, D.C.: Brookings, 1977. Major work on the subject, but should be read critically.

Paul, Diane B. *The Politics of the Property Tax.* Lexington, Mass.: Lexington Books, 1975. Urban focus.

Pechman, Joseph, and Okner, Benjamin. *Who Bears the Tax Burden?* Washington, D.C.: Brookings, 1974. Raises a good question but is flawed by the use of existing assessments as the data base.

People Before Property: A Real Estate Primer and Research Guide. Cambridge, Mass.: Urban Planning Aid, 1972. Guide for doing research on individuals and businesses concerned with property.

Peterson, George E., ed. *Property Tax Reform.* Washington, D.C.: Urban Institute, 1973. Useful summary of the key issues.

Tax Institute of America. *The Property Tax: Problems and Potentials.* Symposium conducted by the Tax Institute of America, November 1966. Princeton, N.J., 1967. Dated but useful summary of the key actors and issues.

USDA, ERS. *Alternative Sources of Local Tax Revenue in Appalachia.* Washington, D.C., 1974. Concludes that the alternatives would not greatly change the present situation in Appalachia.

2. COAL LANDOWNERSHIP AND PROPERTY TAXATION

A. Appalachian Regional and State Studies. There is no comprehensive survey of landownership or property-tax patterns in Appalachia. Many of the studies are unpublished and are not in general circulation. Most of the works listed in this section can be found in the library collection of the Highlander Research and Education Center, New Market, Tennessee.

Akintola, Jacob; Colyer, Dale; and Weber, Wayne. *Rural Land Use in the Monongahela River Basin.* Bulletin 641. Morgantown: WVU College of Agriculture and Forestry, Agricultural Experiment Station, August 1975. Study of ten county area in West Virginia for trends in ownership of land and mineral rights, agricultural production, and limits to expansion.

Barkus, Gary. "The West Virginia Tax Structure, the People and Coal: An Analysis for the Layman." *ARDF Public Interest Report No. 8.* Charleston, W. Va.: Appalachian Research and Defense Fund, 1971. Concludes West Virginia's property tax system is one of the most regressive in the nation.

Blizzard, William. "West Virginia Wonderland." *Appalachian South* 1 (Spring and Summer 1966): 8–15. Also in *MLW* (November 1970): 5-11. Examines the relationship between corporate control and the tax structure.

Childers, Joey. "Absentee Ownership of Harlan County." In F, pp. 81–92. History of land acquisition in Harlan County and an assessment of the situation today.

"County Mirrors Appalachian Patterns: Inequities in the Tax System." *Sandy New Era* (1 February 1979): 4–5. Also in F, pp. 106-07. Information on the Mingo County, West Virginia, tax structure.

Fineman, Howard. "Owners of State's Coal Changing as Energy Firms Move In." Louisville *Courier-Journal* (18 December 1977). Lists top twenty-five owners of coal acreage in Kentucky.

Frazier, Jack. *West Virginia Green.* Part 1, Huntington, W.Va.: Solar Age Press, 1976. Lists top fifty corporations in the state and their land holdings.

Gaventa, John. "Property Taxation of Coal in Central Appalachia." Report for the Senate Subcommittee on Intergovernmental Relations from Save Our Cumberland Mountains, Inc. (SOCM), 1973. Published in F, pp. 76-80. Surveys landownership and coal taxation studies in Kentucky, Tennessee, Virginia, and West Virginia.

Gaventa, John; Ormond, Ellen; and Thompson, Bob. "Coal, Taxation and Tennessee Royalists." Nashville, Tenn.: Vanderbilt Student Health Coalition, 1971 (Unpublished). Survey of ownership and taxation in five coal counties, exposing vast underassessment of coal reserves and corporate concentration in the coalfields.

Kirby, Richard M. "Kentucky Coal: Owners, Taxes, Profits. A Study in Representation Without Taxation." Prepared for the Appalachian Volunteers, 1969. See excerpt in *Appalachian Lookout* 1 (October 1969): 19-27. One of the first exposés of undertaxation of coal reserves in eastern Kentucky.

Leistritz, Larry, and Voelker, Stanley. "Coal Resource Ownership: Patterns, Problems and Suggested Solutions." *Natural Resources Journal* 15 (October 1975). Discusses ownership patterns and examines alternatives.

Lincoln Citizens for Tax Reform. "Who Owns Lincoln County?" Pamphlet. Griffithsville, W.Va. 1978. Also in F, pp. 104-05. Citizen pamphlet on ownership and taxation and their effects on local services.

McAteer, J. Davitt. *Coal Mine Health and Safety: The Case of West Virginia.* N.Y.: Praeger, 1973. One of the first studies of West Virginia ownership. Provides figures for fourteen major coal-producing counties.

McDonald, E. Dandridge, and Kalis, Peter J. "Public Schools and Assessment of Mineral Reserves for Tax Purposes." *ARDF Public Interest Report No. 10.* Charleston, W.Va: Appalachian Research and Defense Fund, 1973. Discovers substantial underassessment of mineral property in four West Virginia counties.

McDowell County Committee for Fair Taxation. *Who Owns McDowell County?* Welch, W.Va., 1980. Citizen pamphlet on ownership and taxation and their effects on local services.

Miller, Tom D. *Who Owns West Virginia?* Huntington, W.Va.: The Huntington Publishing Co., 1975. County-by-county summaries.

Millstone, James C. "East Kentucky Coal Makes Profits for Owners, Not Region." In *Appalachia in the Sixties.* Ed. David Walls and John Stephenson. Lexington: Univ. Press of Kentucky, 1972. Examines how coal owners do not pay their fair share of taxes.

"Ownership vs. Stewardship of Land in Nine Counties of Southern West Virginia." Pamphlet. Catholic Diocese of Wheeling, W.Va., n.d. Review of ownership and poverty patterns with theological and economic arguments about the impact of land concentration.

Privratsky, Bruce, and Randolph, Jane. "Coal Taxes in Southwest Virginia." Unpublished report for the U.S. Senate Subcommittee on Intergovernmental Relations from the Concerned Citizens for Fair Taxes, 1973. Documents major inequities in the tax structure.

Ridgeway, James. *The Last Play: The Struggle to Monopolize the World's Energy Resources.* N.Y. Dutton, 1973. Includes a discussion of coal ownership in Appalachia and profile of major energy companies.

Schommer, Carol. "A Critique of Virginia's Mineral Taxation Program." Unpublished study prepared for consideration by Virginia's Dept. of Taxation and the House of Delegates' Finance Committee from Virginia Citizens for Better Reclamation (VCBR), 1978. Also in F, pp. 95-103. Coal tax assessments and landownership in seven coal counties.

Scroggins, James, and Tudor, Dean. "Report on Mineral Taxation in Tennessee." Unpublished report prepared for the Vanderbilt Student Health Coalition, 1973. Examines methods and results of mineral taxation.

Shamsudin, Mohd. Noor Bin, and Colyer, Dale. *Mineral Rights and Property Taxation in West Virginia.* R.M. No. 74. Morgantown: WVU College of Agriculture and Forestry, Agricultural Experiment Station, July 1979. Examines the reappraisal program in West Virginia and concludes that per acre taxes are still too low.

Walls, David S.; Billings, Dwight B.; Payne, Mary P.; and Childers, Joe F., Jr. "Coal Land and Mineral Ownership." In *A Baseline Assessment of Coal Industry Structure in the ORBES Region.* Prepared for Ohio River Basin Energy Study Region (ORBES). Washington, D.C.: Office of Research and Development, U.S. Environmental Protection Agency, 1979. A comprehensive survey of landownership studies.

Wells, John C., Jr. "Poverty Amidst Riches: Why People Are Poor in Appalachia." Ph.D. dissertation, Rutgers Univ., 1977. See especially pp. 153-98. Excellent review of ownership studies throughout the region.

Impact Studies. Landownership and taxation patterns affect every facet of life in Appalachia, yet few studies address this question in a comprehensive manner. These studies illustrate the various impact areas. No attempt is made to provide exhaustive coverage for any one area. Several studies that trace the evolution of coal landownership patterns are included in this section.

Appalachian Research and Defense Fund. *Coal Government of Appalachia.* Charleston, W. Va. 1972. Documents the overwhelming political and economic influence of the coal industry in West Virginia.

Arnett, Douglas O. "Eastern Kentucky: The Politics of Dependency and Underdevelopment." Ph.D. dissertation, Duke Univ., 1978. Examines the dependency that results from corporate control.

Balliet, Lee. *"A Pleasing Tho' Dreadful Sight": Social and Economic Impacts of Coal Production in the Eastern Coalfields.* Prepared for the Office of Technology Assessment, 1978. Case studies of six counties affected by coal production. Identifies absentee corporate ownership as a principal cause of underdevelopment.

Barkan, Barry, and Baldwin, Lloyd R. "Picking Poverty's Pocket." *MLW* (September 1970): 4-9, 19-21. Impact of absentee ownership in southwest Virginia.

Bethell, Thomas. *The Hurricane Creek Massacre.* N.Y.: Perennial Library, 1972. Inquiry into a coal-mine explosion that illustrates the influence of the coal industry on federal and state coal-safety policy.

Bethell, Thomas, and McAteer, J. Davitt. "The Pittston Mentality: Manslaughter on Buffalo Creek." *Washington Monthly* (May 1972): 19-28. Also in LJ&A, pp. 259-75. Documents how the negligence and indifference of one of Appalachia's largest coal companies led to the Buffalo Creek disaster, which killed over a hundred local residents.

Caudill, Harry. *My Land Is Dying.* N.Y.: Dutton, 1971. Documents the devastation of the Appalachian mountains by strip-miners and corporate feudalism.

———. *Night Comes to the Cumberlands: A Biography of a Depressed Area.* Boston: Atlantic Monthly Press, 1962. How the people of the eastern Kentucky coalfields have suffered under absentee ownership.

Charles River Associates, Inc. *The Economic Impact of Public Policy on the Appalachian Coal Industry and the Regional Economy.* 3 vols. Prepared for the ARC. Washington, D.C., 1973.

Coal Company Monitoring Project. *"... in the mines, in the mines, in the Blue Diamond Mines ...".* Knoxville, Tenn., 1979. Study of one coal company's impact on several Kentucky and Tennessee communities.

Dials, George, and Moore, Elizabeth C. "The Cost of Coal: We Can Afford to Do Better." *Appalachia* 8 (October-November 1974): 1-29. Reprinted from *Environment* (September 1974). Examines the environmental and human costs of coal mining and shows that such costs are unwarranted.

Diehl, Richard. "Appalachia's Energy Elite: A Wing of Imperialism." *Peoples Appalachia* 1 (March 1970): 2-3. Comments on how the energy elite has come to dominate not only the Appalachian, but also the American, political and social systems.

———. "How International Energy Elite Rules." *Peoples Appalachia* 1 (April-May 1970): 1, 7-12. Examines how the energy elite is organized and rules in Appalachia.

Dix, Keith. "Appalachia: Third World Pillage." *Peoples Appalachia* 1 (August-September 1970): 9-13. Compares the absentee exploitation of Appalachia with that in Third World, nonsocialist, underdeveloped countries.

———. "The West Virginia Economy: Notes for a Radical Base Study." *Peoples Appalachia* 1 (April-May 1970): 3-7. Brief historical survey of West Virginia's role as a resource supplier to the rest of the nation, and an analysis of the changing structure of West Virginia's post–World War II economy.

Drake, Richard. "Documents Relating to the Broad Form Deed." *Appalachian Notes* 2:1 (1974): 1-6. The broad-form deed was the document used by coal speculators to acquire extensive mineral rights in the late 1800s and early 1900s.

Egerton, John. "The King Coal Good Times Blues." *New Times* (2 February 1978): 26-34. Shows how the quality of life in the coalfields has not improved significantly for many residents despite the recent coal boom.

Eller, Ronald D. "The Coal Barons of the Appalachian South, 1880-1920." *Appalachian Journal* 4 (Spring-Summer 1977): 195-207. Examines the social attitudes of coal owners-operators in the Appalachian coalfields.

———. "Industrialization and Social Change in Appalachia: A Look at the Static Image, 1880-1930." In LJ&A, pp. 36-46. Explains how the persistent poverty of Appalachia has resulted from the particular kind of industrialization that unfolded in the coalfields from 1880 to 1930.

Gaventa, John. "The Amax Record Elsewhere." A study prepared for the Concerned Citizens of Piney by Save Our Cumberland Mountains (SOCM), 1976 (unpublished). Details effects of Amax Coal Company's actions in states where it mines coal.

———. "Land Ownership and Coal Productivity." In Helen M. Lewis et al., *Coal Productivity and Community: The Impact of the National Energy Plan in the Eastern Coalfields.* Prepared for the U.S. Dept. of Energy, 1978. Also in F, pp. 108-18. Examines the effect of ownership on what coal should be mined by whom and on social and economic conditions in coal communities.

———. *Power and Powerlessness: Quiescence and Rebellion in an Appalachian Valley.* Oxford: Clarendon Press; Urbana: Univ. of Illinois Press, 1980. Examines power relationships in Middlesboro, Kentucky, and surrounding rural areas.

Hardt, Jerry. *Harlan County Flood Report.* Corbin, Ky.: Appalachia—Science in the Public Interest, 1978. Offers evidence that recent flooding was aggravated by strip-mining.

Harlan Miners Speak: Report on Terrorism in the Kentucky Coal Fields. Prepared by members of the National Committee for the Defense of Political Prisoners. N.Y.: Harcourt, 1932; repr. N.Y.: DeCapo, 1970. Includes a chapter, "Who Owns the Mines." Describes the power and influence of the coal industry in the later 1920s and early 1930s.

Harvey, Curtis, et al. *Coal and the Social Sciences: A Bibliographical Guide to the Literature.* Lexington: Social Science/Technology Development Group, Univ. of Kentucky, 1979.

"Housing Crisis in the Coalfields." *United Mine Workers Journal* (16-29 February 1976). Series of articles that demonstrate the impact of absentee ownership and corporate control on housing in the coalfields.

"Housing in Appalachia." Part 1, *MLW* (January 1979): 3-28 and Part 2, *MLW* (February 1979): 20-27. Good survey of housing problems in the coalfields.

Kaufman, Paul. "Poor Rich West Virginia." *New Republic* (2 December 1972): 12-15. Also in F, pp. 126-28. Illustrates the effect of absentee ownership in West Virginia on the tax structure and the people.

Landy, Marc K. *The Politics of Environmental Reform: Controlling Kentucky Strip Mining.* Baltimore: JHP, 1976. Illustrates the continuing influence of the coal industry in state politics.

Lewis, Helen M. *Coal Productivity and Community: The Impact of the National Energy Plan in the Eastern Coalfields.* Prepared for the Dept. of Energy, 1978. Anticipates the effects of increasing coal production on communities in the eastern coalfields and tries to determine the relationship between social impacts on community and coal-miner productivity. Includes community studies from Kentucky, West Virginia, Ohio, Virginia, and Alabama.

Lewis, Helen M.; Johnson, Linda; and Askins, Don, eds. *Colonialism in Modern America: The Appalachian Case.* Boone, N.C.: Appalachian Consortium Press, 1978. Excellent collection of articles documenting the effects of outside corporate control of Appalachia.

Lewis, Helen M.; Kobak, Sue E.; and Johnson, Linda. "Family, Religion and Colonialism in Central Appalachia, or Bury My Rifle at Big Stone Gap." In LJ&A, pp. 113-39. Examines the impact of absentee ownership and control on family and religion in Appalachia.

McAteer, J. Davitt. "You Can't Buy Safety at the Company Store." *Washington Monthly* (November 1972): 7-19. Compares United States coal health and safety practices with those in other countries and concludes that corporate ownership in America is a primary cause of our dismal health and safety record.

Munn, Robert F. *The Coal Industry in America: A Bibliography and Guide to Studies.* Morgantown: WVU Library, 1977. Useful introduction to the literature.

Murphy, Thomas. "The Investment Nobody Knows About." *Dun's Review* (April 1965): 40-43, 131-32. Discusses the profits from coal royalties that come from leasing mineral-rich lands to mining companies.

Murray, Francis X. *Where We Agree: Report of the National Coal Policy Project.* 2 vols. Boulder, Colo.: Westview, 1978. Discusses problems associated with the coal industry and offers recommendations.

National Sacrifice Area. Williamson, WV.: Appalachian Alliance, 1979. Information on the impact of absentee ownership on taxation, housing, and health.

Noyes, R., ed. *Coal Resources, Characteristics and Ownership in the USA.* Park Ridge, N.J.: Noyes Data Corporation, 1968. Useful statistical source.

"Ralph Nader Letter." In LJ&A, pp. 71-83. Also in F, pp. 119-25. Letter written by Nader in 1973 to the chairman of the board of the London-based American Association, Ltd., detailing the exploitative nature of the association's absentee control of 50,000 acres in the Appalachian coalfields.

Appendix Three

Seltzer, Curtis I. "The United Mine Workers of America and the Coal Operators: The Political Economy of Coal in Appalachia." Ph.D. dissertation, Columbia Univ., 1978. Describes and assesses the impact of coal development on the people and resources of the Appalachian coalfields.

Shackelford, Laurel, and Weinberg, Bill. *Our Appalachia: An Oral History.* N.Y.: Hill & Wang, 1977. Includes a section on John Mayo, who, by 1910, had spent over twenty years buying mineral rights and selling them to absentee corporations and other speculators, as well as local reaction to tourist development in North Carolina.

Simon, Richard M. "The Labor Process and Uneven Development in the Appalachian Coalfields." *International Journal of Urban and Regional Research.* 4(March 1980):46-71. Focuses on West Virginia in an effort to develop a model of regional underdevelopment.

Smith, Janet M.; Ostendorf, David; and Schechtman, Mike. *Who's Mining the Farm?* Herrin, Ill.: Illinois South Project, 1978. Though outside Appalachia, this study provides an excellent example of the impact of coal landownership on agricultural communities.

Sobek, Andrew A., and Streib, Donald. *A Selective Bibliography of Surface Coal Mining and Reclamation Literature.* Vol. 1: *Eastern Coal Province.* Prepared for the U.S. Dept. of Energy as a part of the Argonne Land Reclamation Program. Springfield, Va.: National Technical Information Service, 1977. One of the more recent surveys of this literature.

Tompkins, Dorothy Campbell. *Strip Mining for Coal.* Berkeley: Institute of Governmental Studies, Univ. of California, 1973. Bibliography.

U.S. President's Commission on Coal. *The American Coal Miner: A Report on Community and Living Conditions in the Coalfields.* Washington, D.C.: USGPO, 1980. Principal volume of the commission's report. Downplays the impact of absentee corporate ownership.

Walls, David. "Central Appalachia in Advanced Capitalism: Its Coal Industry Structure and Coal Operator Associations." Ph.D. dissertation, Univ. of Kentucky, 1978. Provides an important theoretical perspective along with useful information on the coal industry.

Walls, David; Billings, Dwight; Payne, Mary; and Childers, Joe. *A Baseline Assessment of Coal Industry Structure in the Ohio River Basin Energy Study Region.* Washington, D.C. Office of Research and Development, U.S. Environmental Protection Agency, 1979. Includes detailed information on coal production and employment by county, coal land- and mineral ownership, and projections of coal development.

Weller, Jack. "Appalachia: America's Mineral Colony." *Vantage Point,* No. 2, n.d. [about 1974]. Also in LJ&A, pp. 47-55. Describes the various effects of outside ownership.

Williams, John A. *West Virginia and the Captains of Industry.* Morgantown: WVU Library, 1976. Discusses West Virginia history from 1880 to 1913 in terms of four corporate leaders who were instrumental in developing and maintaining a "colonial political economy" in the state.

Wright, Warren. "The Big Steal." In LJ&A, pp. 161-75. Study of legal maneuvers through which companies obtained control of east Kentucky land.

Young, Robert A., and Stepko, George, Jr. *Ownership and Land Use Constraints upon the Recoverability of Coal: A Methodology and Test Case.* Morgantown: WVU, 1975. Uses Monongalia County, West Virginia, as a study area.

3. RECREATION LANDOWNERSHIP AND SECOND-HOME DEVELOPMENT

Alanen, A. R., and Smith, K. E. "Growth versus No-Growth Issues, with an American Appalachian Perspective." *Tijdschrift voor Economische en Sociale Geografie* 68:1 (1977): 30-42. Pinpoints sources of growth and no-growth sentiments in Pocahontas, West Virginia, with specific reference to recreational development.

Allan, Leslie; Kudar, Beryl; and Oakes, Sarah L. *Promised Lands.* Vol. 1: *Subdivisions in Deserts and Mountains.* N.Y.: INFORM, 1976. Includes chapters on the land-subdivision industry and on the impact of subdivisions on the mountain environment.

American Society of Planning Officials. *Recreational Lot and Second Home Development: A Manual for Reviewing Impacts.* Washington, D.C.: Council on Environmental Quality, 1976. Handbook for government officials and planners.

―――. *Subdividing Rural America: Impacts of Recreational Lot and Second Home Development.* Washington, D.C.: USGPO, 1976. Concludes there is a potential for significant adverse impacts from recreational and second-home developments. Useful bibliography.

Appalachia Business Review 2 (Fall 1973). Issue devoted to industrial location and recreation in western North Carolina.

ARC. *State and Regional Development Plans in Appalachia.* Washington, D.C., 1968. Describes the Appalachian Highland region as having "its greatest potential as a recreation, tourism and resource area."

Bingham, Edgar. "The Impact of Recreational Development on Pioneer Life Styles in Southern Appalachia." *Proceedings of the Pioneer America Society,* 2 (1973): 59-68. Also in LJ&A, pp. 57-69. Examines the effect of recreational development on Appalachian culture and suggests possible approaches for preserving the culture.

Binkley, Clark, and Yale School of Forestry and Environmental Studies. *Estimating Recreation Benefits: A Critical Review and Bibliography.* Exchange Bibliography 1219. Monticello, Ill.: Council of Planning Librarians, February 1977. Concerned with quantifying the dollar value that individuals ascribe to a new or expanded facility.

Bosselman, Fred P. *In the Wake of the Tourist: Managing Special Places in Eight Countries.* Washington, D.C.: Conservation Foundation, 1978. Examines many facets of tourism and its impact on environmentally sensitive areas in eight countries.

Bosselman, Fred P. and Callies, David. *The Quiet Revolution in Land Use Control.* Prepared for the Council on Environmental Quality. Washington, D.C.: USGPO, 1971. Useful discussion of recreational land-use control.

Appendix Three   195

Branscome, Jim, and Matthews, Peggy. "Selling the Mountains." *Southern Exposure* 2 (Fall 1974): 122-29. Also in F, pp. 144-51. Presents the case against recreational and second-home development.

Brewster, Lawrence F. *Summer Migrations and Resorts of South Carolina Low-Country Planters.* Durham: Duke Univ. Press, 1947; repr. N.Y.: AMS Press, 1970. Study of some of the first Appalachian tourists in the early 1800s.

Brown, Richard N., Jr. *Economic Impact of Second-Home Communities: A Case Study of Lake Latanka, Pa.* ERS-452. Washington, D.C.: ERS, USDA, 1970. Analysis of the economic impacts generated by one of many second-home communities in the Northeast.

Carey, Omer L. "The Economics of Recreation: Progress and Problems." *Western Economics Journal* 3 (Spring 1965): 172-81. Examines the various approaches to evaluating the economic benefits of recreation.

Cary, William; Johnson, Molly; Golden, Meredith; and Van Noppen, Trip. *The Impact of Recreational Development. A Study of Land Ownership, Recreational Development and Land Use Planning in the North Carolina Mountain Region.* Durham: North Carolina Public Interest Research Group, 1975. Ownership patterns in ten North Carolina mountain counties.

Chen, David Y. *The Seasonal Tourist Accommodation Industry in Western North Carolina: A Report to Resort Owners/Operators.* Greensboro: North Carolina A & T State Univ., 1976.

Clawson, Marion, and Knetsch, Jack L. *Economics of Outdoor Recreation.* Published for RFF. Baltimore: JHP, 1967. Examines the economic and social aspects of outdoor recreation.

Coppock, J. T., ed. *Second Homes: Curse or Blessing?* Oxford: Pergamon Press, 1977. Offers an international perspective.

Cornett, Judy. "A Preliminary Bibliography of Materials Relating to Recreation and Second-Home Development in the Southern Appalachians." Prepared for the Research and Public Policy Committee of the Appalachian Studies Conference, 1979 (unpublished). A useful bibliography from which some of the sources and annotations in this section are taken.

Cutler, M. Rupert. "The Tragic Story of Magic Mountain." *Living Wilderness* 29 (Summer 1965): 7-9. A parable (of thinly disguised Spruce Knob, West Virginia) of the overdevelopment of a once-natural scenic area.

Dobson, Jerome E. "The Changing Control of Economic Activity in the Gatlinburg, Tennessee, Area, 1930-1973." Ph.D. dissertation, Univ. of Tennessee, 1975. Concludes that "indigenous people are participating less and less in the major benefits of Gatlinburg's economic growth."

Godschalk, David R.; Parker, Francis H.; and Roe, Charles E., eds. *Land Development in the North Carolina Mountains: Impact and Policy in Avery and Watauga Counties.* Chapel Hill: Dept. of City and Regional Planning, Univ. of North Carolina, June 1975. Documents the impact of recreational land development on the quality of life in two Appalachian counties.

Goeldner, C. R., and Dicke, Karen. *Bibliography of Tourism and Travel. Research Studies, Reports, and Articles.* 3 vols. Boulder: Graduate School of Business Administration, Univ. of Colorado, 1971.

Gottfried, Robin. "Observations on Recreation-Led Growth in Appalachia." *American Economist* 21 (Spring 1977): 44-50. Examines the pros and cons of recreation development in Appalachia and concludes that serious questions remain as to its benefits.

Hansen, David E., and Dickinson, Thomas E. "Undivided Interests: Implications of a New Approach to Recreational Land Development." *Land Economics* 51 (May 1975): 124-32. Offers an alternative method of developing and marketing recreational properties.

Johns, Irwin R., and Smith, Norman G. *A Profile of Non-Resident Recreation Property Owners in an Appalachian County.* MP-589. Maryland Agricultural Experiment Station, March 1967. Focus on the attraction of rural areas to seasonal residents.

Johnson, Hugh A.; Carpenter, J. Raymond; and Dill, Henry W., Jr. *Exurban Development in Selected Areas of the Appalachian Mountains.* ERS-111. Washington, D.C.: ERS, USDA, April 1963. Focus on northwestern Virginia and adjacent areas of West Virginia.

Johnson, Hugh A.; Huff, Judith M.; and Csorba, J. J. *Private Outdoor Recreation Enterprises in Rural Appalachia.* ERS-429. Washington, D.C.: ERS, USDA, 1969. Seeks to determine the extent to which recreation enterprises operating in Appalachia can help meet the growing urban demands for outdoor recreation and provide profitable use of rural resources and employment for rural people.

Klain, Ambrose, and Phelan, Dennis M. *Second Homes, Vacation Homes: Potentials, Impacts, and Issues: An Annotated Bibliography.* Exchange Bibliography 839. Monticello, Ill.: Council of Planning Librarians, 1975.

Lamm, Joy. "So You Want a Land-Use Bill? The Case of the North Carolina Mountain Area Management Act." *Southern Exposure* 2 (Fall 1974): 52-62. Illustrates the influence of the real-estate lobby.

Little, Arthur D., Inc. *Tourism and Recreation: A State-of-the-Art Study.* Prepared for Office of Regional Development Planning. Washington, D.C.: U.S. Economic Development Administration, 1967. Includes a section on the economic impact on a regional or local economy. Bibliography.

MacCannell, Dean, and Meyers, Phyllis. *So Goes Vermont.* Washington, D.C.: Conservation Foundation, 1974. Vermont has experienced recreation and land-use problems similar to those in Appalachia.

Morris, John. "The Potential of Tourism." *The Southern Appalachian Region: A Survey.* Ed. Thomas Ford. Lexington: Univ. of Kentucky Press, 1962.

Nathan [Robert R.] Associates. *Recreation as an Industry.* 2 vols. Prepared for the ARC and issued as Research Report No. 2. Washington, D.C., 1966. Outlines the advantages of recreation in economic development, but points out that most jobs connected with the industry are seasonal and seldom "pay a living wage."

O'Neill, Frank. "Greatest Menace Yet to Southern Mountains." *Southern Voices* 1 (May-June 1974): 73-78. Land development in Georgia.

Osborn, Howard A. *Spruce Knob–Seneca Rocks National Recreation Area: Impact on the Local Community.* Washington, D.C.: ERS, USDA, May 1976. Examines the income effects of a national recreation area on three West Virginia counties.

Parlow, Anita. *The Land Development Rag: The Impact of Resort Development on Two Appalachian Counties, Watauga and Avery in North Carolina.* Knoxville: Southern Appalachian Ministry in Higher Education, 1976. Excerpt in LJ&A, pp. 177-98.

Payne, Brian R.; Gannon, Richard; and Ireland, Lloyd Co. *The Second-Home Recreation Market in the Northeast: A Problem Analysis of Economic, Social and Environmental Impacts.* Washington, D.C.: USGPO, 1975. Documents the various impacts of the second-home recreation market and suggests policy responses and a program of research.

Ragatz [Richard L.] Associates. *Future Demands for Recreational Properties.* Washington, D.C.: Urban Land Institute, 1974. Analysis of factors that will influence recreational development for the next twenty years.

———. *Recreational Properties in Appalachia: An Analysis of the Markets for Privately Owned Recreational Lots and Leisure Homes.* Prepared for the ARC. Springfield, Va.: National Technical Information Service, 1974.

"Resort Fever is Changing Face of Appalachia." *U.S. News & World Report* (14 April, 1980): 58-61. Examines the impact of land developers on Appalachia.

"Rural Impact of Recreational Development." In *Planning Frontiers in Rural America.* Papers and Proceedings of the Boone Conference, 16-18 March 1975. Washington, D.C.: USGPO, 1976. Articles focusing on east Tennessee and West Virginia

Shands, William, E., and Woodson, Patricia. *The Subdivision of Virginia's Mountains: The Environmental Impact of Recreational Subdivisions in the Massanutten Mountain-Blue Ridge Area, Virginia.* Washington, D.C.: Conservation Foundation, 1974.

Sinclair, Robert O. & Meyer, Stephen B. *Non-Resident Ownership of Property in Vermont,* Bulletin 670. Vermont Agricultural Experiment Station, 1972.

Thomas, Dana L. *Lords of the Land: The Triumphs and Scandals of America's Real Estate Barons—from Early Times to the Present.* N.Y.: Putnam, 1977. Includes chapters on the Founding Fathers, Astors, railroads, and lucrative farm acres.

Urban Land Institute. *Land: Recreation and Leisure.* Abstracted from the First Annual Land Use Symposium. Washington, D.C., 1970. Chap. 12 details the growth of Beech Mountain, Banner Elk, North Carolina. Good Bibliography.

URS Research Co. *Recreation Potential in the Appalachian Highlands: A Market Analysis.* Prepared for the ARC. Washington, D.C., 1971. Argues for increased recreation development in Appalachia.

U.S. Department of Interior, Bureau of Outdoor Recreation. *A Report on Outdoor Recreation Demand, Supply and Needs in Appalachia.* Prepared for the ARC. Washington, D.C., 1967. Gives agency view of positive values of recreation.

Walp, Neil. "The Market for Recreation in the Appalachian Highlands." *Appalachia* 4 (November-December 1970): 27-36. Describes twenty-three recreation complexes in Appalachia and discusses the potential recreational market in the region.

Whisnant, David. *Modernizing the Mountaineer: People, Power, and Planning in Appalachia.* N.Y.: Burt Franklin, 1980. Includes a discussion on the ARC's recreation and tourism programs.

4. FARM LANDOWNERSHIP

Farm Landownership and Taxation. This section includes works on rural landownership at a regional, state, and local level. Also includes works analyzing the foreign ownership of rural land, black landownership, and taxation of agricultural land. Works on landownership at the national level are listed in Section 1.

Atkinson, Glen W. "The Effectiveness of Differential Assessment of Agricultural and Open Space Land." *American Journal of Economics and Sociology* 36 (April 1977): 197-204. Finds the current laws have little impact on land use.

Barlowe, Raleigh, and Timmons, John F. *Farm Ownership in the Midwest.* Research Bulletin 361. Iowa Agricultural Experiment Station, June 1949. Examines who owns the farms in the midwest, how they were acquired, and the form in which they are held.

Bird, Richard M. *Taxing Agricultural Land in Developing Countries.* Cambridge: Harvard Univ. Press, 1974. Provides insights relevant to the American situation.

Boles, Donald E., and Rupnow, Gary. "Local Governmental Functions Affected By the Growth of Corporate Agricultural Land Ownership: A Bibliographic Review." *Western Political Quarterly* 32 (December 1979): 467-78.

Browne, Robert S. *Only Six Million Acres: The Decline of Black Owned Land in the Rural South.* N.Y.: Black Economic Research Center, 1973. Examines the major decline in black-owned land.

Buckeye Hills Resource Conservation and Development Project. *Land Ownership Study 1976: Belmont-Monroe-Morgan-Noble-Washington Counties, Ohio.* Caldwell, Ohio, 1976. Documents the presence of a large number of nonresident landowners.

Carlin, Thomas A., and Woods, W. Fred. *Tax Loss Farming.* ERS-546. Washington, D.C.: ERS, USDA, April 1974. Examines the complexities of the issue.

Conklin, H. E., and Bryant, W. R. "Agricultural Districts: A Compromise Approach to Agriculture Preservation." *American Journal of Agricultural Economics* 56 (August 1974): 607-13. Examines New York's experience with agricultural districts as a means of preserving farmland.

Cotner, Melvin L. *Land Use Policy and Agriculture: A State and Local Perspective.* ERS-650. Washington, D.C.: ERS, USDA, February 1977. Argues that America is not running out of farmland.

Crowe, Kenneth, C. *America for Sale.* Garden City, N.Y.: Doubleday, 1978. Journalistic account of the recent surge of foreign investments in United States land.

Dangerfield, Jeanne. "Sowing the Till: A Background Paper on Tax Loss Farming." *Congressional Record* 119 (16 May 1973): 9247-55. Sets forth the arguments against tax-loss farming.

Dorgan, Byron, *The Progressive Land Tax: A Tax Incentive for the Family Farm.* Washington, D.C.: Conference on Alternative State and local Policies, 1978. One of the best of a number of studies on property tax reform that would benefit the family farmer.

Appendix Three 199

Gardner, Delworth. "The Economics of Agricultural Land Preservation." *American Journal of Agricultural Economics* 59 (December 1977): 1027-36. Concludes that agricultural land-retention legislation is the wrong thing at the wrong time. Contends that it has not been adequately demonstrated that more American land than the market will make available will be needed to produce food in the next decade.

Gloudemans, Robert J. *Use-Value Farmland Assessments: Theory, Practice and Impact.* Chicago: IAAO, 1974. In-depth analysis of the key issues.

Hady, Thomas, F., and Sibold, Ann G. *State Programs for the Differential Assessment of Farm and Open Space Land.* AER-256. Washington, D.C.: ERS, USDA, April 1974. Short discussion of issues followed by a complete listing and description of state statistics as of 1973.

Hoffsommer, Harold, ed. *The Social and Economic Significance of Land Tenure in the Southwestern States.* A Report of the Regional Land Tenure Research Project. Chapel Hill: Univ. of North Carolina Press, 1950. Measures the relationship between the tenure status of the family farm and its economic and social performance.

Inman, Buis T., and Fippin, William H. *Farm Land Ownership in the United States.* MP-699. Washington, D.C.: USDA, December 1949. Includes land in farms by major types of owners and extent of holdings by individual owners.

IAAO. *Property Tax Incentives for Preservation: Use-Value Assessment and the Preservation of Farmland, Open Space, and Historic Sites.* Proceedings of the 1975 Property Tax Forum. Chicago, 1975. Nine essays examine the rationale for existing use-value assessment programs as well as their effectiveness in achieving their objectives.

Johnson, Bruce B. *Farmland Tenure Patterns in the United States.* AER-249. Washington, D.C.: ERS, USDA, February 1974. Detailed examination of farmland ownership and control based on 1969 Census data.

Keene, John. *Untaxing Open Space: An Evaluation of the Effectiveness of Differential Assessment of Farms and Open Space.* Prepared for the Council on Environmental Quality. Washington, D.C.: USGPO, 1976. Study of forty-two state laws that give preferential tax treatment to agricultural land.

Kolesar, John, and Scholl, Jaye. *Misplaced Hopes, Misspent Millions: A Report on Farmland Assessment in New Jersey.* Princeton, N.J.: Center for Analysis of Public Issues, 1972. Highly critical of use-value assessment. Estimates a tax loss of $48 million per year with no return. Contends that outright speculators control 10 percent of the land in the program.

―――. *Saving Farmland.* Princeton, N.J.: Center for Analysis of Public Issues, 1975. Reexamines techniques for the preservation of agricultural land. Criticizes a plan to purchase development rights of farmland.

Krause, Kenneth R. *Foreign Investment in the U.S. Food and Agricultural System: An Overview.* AER-456. Washington, D.C.: ESCS, USDA, May 1980. Concludes that foreign investors do not have dominant control of firms throughout the input, production, marketing, and processing of any one food item.

"Land Ownership Patterns and Community Development." Pamphlet prepared by the Alabama Agricultural Marketing Project, n.d. Compares land patterns and community development indicators in twenty Alabama counties.

Lewis, James A. *White and Minority Small Farm Operators in the South.* AER-353. Washington, D.C.: ERS, USDA, 1976. Identifies, compares, and contrasts resources and characteristics of small farm operators in thirteen Southern states.

McGee, Leo, and Boone, Robert, eds. *The Black Rural Landowner—Endangered Species: Social, Political and Economic Implications.* Westport, Conn.: Greenwood, 1979. Important collection of essays on the black farmer.

Miller, Judy K. "Where Does All of Our Farmland Go?" *Mountain Review* 4 (January 1979): 35-38. General discussion of the forces leading to the loss of farmland in Appalachia.

Ognibene, Peter J. "Vanishing Farmlands: Selling Out the Soil." *Saturday Review* (May 1980): 29-32.

Roberts, Neal A., and Brown, H. James, eds. *Property Tax Preferences for Agricultural Land.* Montclair, N.J.: Allanheld, Osmun, 1980. Helpful analysis of the various types of property-tax preferences.

Salamon, Lester M. *Land and Minority Enterprise: The Crisis and the Opportunity.* Washington, D.C.: U.S. Dept. of Commerce, 1976. Includes a detailed study of Black land loss in Southern states, an examination of the long-term effects of landownership on rural Black families, and an analysis of the potential uses of publicly owned land for minority economic development.

Southern, John H. *Farm Land Ownership in the Southwest.* Bulletin 502. Arkansas Agricultural Experiment Station, December 1950. Information on ownership for five Southwestern states.

Stamm, Jerome M., and Sibold, Ann G. *Agriculture and the Property Tax: A Forward Look Based on a Historical Perspective.* AER-392. Washington, D.C.: ERS, USDA, November 1977. Evaluates the importance of the property tax to the agricultural sector.

Strohbehn, Roger W. *Ownership of Rural Land in the Southeast.* AER-46. Washington, D.C.: ERS, USDA, December 1963. Examines who owns the rural land and how it is used in seven Southeastern states.

Strohbehn, Roger, W., and Wunderlich, Gene. *Land Ownership in the Great Plains States, 1958: A Statistical Summary.* SB-261. Washington, D.C.: USDA, April 1960. Examines characteristics of landowners and trends in ownership patterns in the ten Great Plains states.

Tharp, Max M. *Farm Land Ownership in the Southeast.* Bulletin 378. South Carolina Agricultural Experiment Station, June 1949. Examines who owns the farms in the Southeast and how these farms were acquired.

Theilacker, J. *A Selected Bibliography on Agricultural Land Preservation.* P-355. Monticello, Ill.: Vance Bibliographies, 1979.

Thompson, Allen R., and Greene, Michael. *The Status of Minority Farms in the U.S., 1974.* Durham, N.H.: Center for Industrial and Institutional Development, Univ. of New Hampshire, April 1980. Examines Black, Indian, Hispanic, and Oriental farm owners.

Turner, Howard. *A Guide to Farm Tenure Data in Census Publication.* Washington, D.C.: Bureau of Agricultural Economics, USDA, May 1948. Reference tool for researchers looking for farm tenure data in agricultural census publications since 1880.

Appendix Three

U.S. Congress, Senate. Committee on Agriculture, Nutrition, and Forestry. *Foreign Investment in United States Agricultural Land.* 95th Congress, 2nd session. Washington, D.C.: USGPO, January 1979. Summary of key issues.

USDA. *Perspectives on Prime Lands.* Background Papers for Seminar on the Retention of Prime Lands, 16-17 July 1975. Washington, D.C., 1975. Background information on prime production land in America.

USDA, ERS. *Farm Real Estate Taxes: Recent Trends and Developments.* Washington, D.C., annually. Data on property taxes paid by farmland owners and on taxes in relation to farm income and sales. Based on nationwide sample reported by state and region.

USDA, ESCS. *Foreign Ownership of U.S. Agricultural Land.* AER-447. Washington, D.C., February 1980. Foreigners own less than 0.5 percent of all privately held agricultural land.

U.S. General Accounting Office. *Foreign Ownership of United States Farmland—Much Concern, Little Data.* Washington, D.C.: USGPO, 1978. Discusses state efforts to control foreign ownership, problems with identifying foreign investment in agriculture, and the impact of foreign investment on land prices.

Walrath, Arthur J. *Rural Land Ownership and Economic Development in a Three-County Area.* Bulletin 10. Blacksburg: VPI Research Division, in cooperation with ERS, USDA, September 1967. Study of Alleghany and Bath Counties in Virginia, and Greenbrier County, West Virginia.

———. *Rural Land Ownership and Industrial Expansion.* Bulletin 527. Blacksburg: VPI Agricultural Experiment Station, in cooperation with ERS, USDA, May 1961. The ownership of all properties of three acres or more in Augusta County, Virginia.

Washington, Harold R. *Black Land Manual: "Got Land Problems?"* Frogmore, S.C.: Penn Community Services, Inc. 1973. Surveys laws of South Carolina as they relate to rights of heirs, condemnation, mortgage foreclosures, etc., in order to help Black people retain more of their land.

Wunderlich, Gene. *Land Along the Blue Ridge: Ownership and Use of Land in Rappahannock County, Virginia.* AER-299. Washington, D.C.: ERS, USDA, July 1975. Results of a 1974 mail survey of landowners.

Family Farm Issues

Aronowitz, Stanley. *Food, Shelter and the American Dream.* N.Y.: Seabury, 1975. Attempts to analyze the forces underlying the food and energy crises.

Bailey, Warren R. *The One-Man Farm.* ERS-519. Washington, D.C.: ERS, USDA, August 1973. Concludes that most of the economies associated with size in farming are achieved by the one-man, fully mechanized farm.

Ball, A. Gordon, and Heady, Earl O., eds. *Size, Structure and Future of Farms.* Ames: Iowa State Univ. Press, 1972. Essays identify the forces that have encouraged the trend toward larger farms and examine the consequences of this trend.

Barrons, Keith C. *The Food in Your Future: Steps to Abundance.* N.Y.: Van Nostrand, 1975. Thoughts of an agricultural technologist regarding the basic aspects of food and agriculture.

Belden, Joe; Edwards, Gibby; Guyer, Cynthia; and Webb, Lee, eds. *New Directions in Farm, Land and Food Policies: A Time for State and Local Action.* Washington, D.C.: Agricultural Project, Conference on Alternative State and Local Policies, 1978. Excellent resource guide to key agricultural issues facing the nation.

Belden, Joe, and Forte, Gregg. *Toward a National Food Policy.* Washington, D.C.: Exploratory Project for Economic Alternatives, 1976. Analysis and critique of federal agricultural policies with broad recommendations for radical change.

Berry, Wendell. *The Unsettling of America: Culture and Agriculture.* San Francisco: Sierra Club, 1977. A powerful attack on agribusiness and a plea to protect the small farmer in order to preserve what is human and humane in our culture.

Bosch, Robert van den. *The Pesticide Conspiracy.* N.Y.: Doubleday, 1978. A telling attack on the pesticide industry and a defense of integrated pest management as an alternative.

Burbach, Roger, and Flynn, Patricia. *Agribusiness in the Americas.* N.Y.: Monthly Review Press, 1980. Examines the rise of corporate agriculture in America and analyzes the global expansion of United States agribusiness transnationals.

Center for Science in the Public Interest. *From the Ground Up: Building a Grass Roots Food Policy.* Washington, D.C., 1976. Action handbook for people who want to reform food policies at the city, county, and state levels.

Clawson, Marion. *Policy Directions for U.S. Agriculture: Long-Range Choices in Farming and Rural Living.* Published for RFF. Baltimore: JHP, 1968. Attempt to present a "comprehensive" view of agricultural policy.

Cochrane, Willard W., and Ryan, Mary E. *American Farm Policy, 1948-1973.* Minneapolis: Univ. of Minnesota Press, 1976. Examines the goals and shortcomings of U.S. farm policy.

Coltrane, R. I., and Baum, E. L. *An Economic Survey of the Appalachian Region with Special Reference to Agriculture.* AER-69. Washington, D.C.: ERS, USDA, April 1965.

Commoner, Barry. *The Poverty of Power.* N.Y.: Knopf, 1976. Good section on agriculture and energy.

Corporate Data Exchange, Inc. *CDE Stock Ownership Directory—Agribusiness: Who Owns and Controls the Nation's Food System.* N.Y., 1978. Comprehensive guide to the largest stockholders of 222 agribusiness corporations, both public and private.

Coughlin, Kenneth M., ed. *Perspectives on the Structure of American Agriculture, Vol. I: The View from the Farm—Special Problems of Minority and Low-Income Farmers.* Washington, D.C.: Rural America, 1980. Emphasis on personal accounts.

———. *Perspectives on the Structure of American Agriculture, Vol. II: Federal Farm Policies—Their Effects on Low-Income Farmers and Rural Communities.* Washington, D.C.: Rural America, 1980. Includes essays on price supports, agricultural research, farm credit, and the encroachment of corporations.

Ebeling, Walter. *The Fruited Plain: The Story of American Agriculture.* Berkeley: Univ. of California Press, 1979. A mainstream history.

Fisher, Steve, and Harnish, Mary. "Losing A Bit of Ourselves: The Decline of the Small Farmer." In *Appalachia/America.* Ed. Wilson Somerville. Boone,

N.C.:Appalachian Consortium Press, 1981. Summarizes the case for the small farmer, the obstacles he faces, and possible reforms.

Fliegel, Frederick C. *The Low-Income Farmer in a Changing Society.* Bulletin 731. Pennsylvania Agricultural Experiment Station, March 1966. Raises questions about the basic nature of low-income farm people and the rural society in which they live.

Fligstein, Neil, et al. *The Transformation of Agriculture in the United States.* Madison, Wisc.: Dept. of Rural Sociology and Program for Class Analysis and Historical Change, 1979 (unpublished manuscript). A collective Marxist work group's attempt to analyze American agriculture.

Fowler, Cary. *Graham Center Seed Directory.* Wadesboro, N.C.: Frank Porter Graham Center, 1979. Includes an important section on the efforts of corporations to gain patent control over seeds.

Friedland, William H., and Barton, Amy. *Destalking the Wily Tomato.* Davis: College of Agricultural and Environmental Sciences, Univ. of California, 1975. Case study of the social consequences of agricultural research in California.

Frundt, Hank. "The Food Gamble." In *U.S. Capitalism In Crisis.* Ed. the Union for Radical Political Economics. N.Y., 1978. Examines how food is used by monopoly capital to manipulate the economic crisis.

Goldschmidt, Walter. *As You Sow: Three Studies in the Social Consequences of Agribusiness.* Glencoe, Ill.: Free Press, 1947; repr. Montclair, N.J.: Allanheld, Osmun & Co., 1978. Classic study of the impact of corporate farming on rural communities.

Goss, Kevin, and Rodefeld, Richard D. *Corporate Farming in the United States: A Guide to Current Literature, 1967-1977.* University Park: Dept. of Agricultural Economics and Rural Sociology, the Pennsylvania State Univ., 1978. Very useful summary of the literature.

Griswold, A. Whitney. *Farming and Democracy.* New Haven: Yale Univ. Press, 1952. Compares farming practices in Britain, France, and the United States. Concludes that the social and economic structure of family farms is more conductive to democratic processes than that of large estates.

Hadwiger, Don F., and Browne, William P., eds. *The New Politics of Food.* Lexington: Lexington Books, 1978. Focus on the politics of agriculture.

Halcrow, Harold G. *Food Policy for America.* N.Y.: McGraw-Hill, 1977. Basic theme is that the most effective progress toward food-policy goals can be obtained only by more complete rural-urban understanding and "a broader participation in a democracy of intellect."

Hamilton, Martha M. *The Great American Grain Robbery and Other Stories.* Washington, D.C.: Agribusiness Accountability Project, 1972. Highly critical of USDA policies.

Hightower, James. "The Case for the Family Farmer." *Washington Monthly* 5 (September 1973): 25-33. Presents a strong, readable case for the small farmer.

———. *Eat Your Heart Out: Food Profiteering in America.* N.Y.: Crown Publishers, 1975. Includes a discussion of the effect of food monopolies on family farmers.

Hightower, Jim, and DeMarco, Susan. *Hard Tomatoes, Hard Times: The Failure of the Land Grant College Complex.* Cambridge, Mass.: Schenkman, 1972. Documents how the land-grant colleges serve the interests of agribusiness to the disadvantage of the family farmer.

Interfaith Center on Corporate Responsibility, ed. *Agribusiness Manual. Background Papers on Corporate Responsibility and Hunger Issues.* N.Y., 1978.

Kramer, Mark. *Three Farms: Making Milk, Meat and Money from the American Soil.* Boston: Atlantic–Little, Brown, 1980. Describes in detail what farmers do and why.

Kravitz, Linda. *Who's Minding the Co-Op? A Report on Farmer Control of Farmer Cooperatives.* Washington, D.C.: Agribusiness Accountability Project, March 1974. Examines what is wrong with the farmer cooperative movement.

Krebs, A. V. *1976 Directory of Major U.S. Corporations Involved in Agribusiness.* San Francisco: Agribusiness Accountability Project, 1976.

Lappe, Frances, M., and Collins, Joseph, with Cary Fowler. *Food First: Beyond the Myth of Scarcity.* Boston: Houghton, 1977. Excellent summary of the hunger issue.

Lenke, Hal. "Land Poor: The Appalachian Farmer." *Peoples Appalachia* 3 (Spring 1973): 35-36.

Lerza, Catherine, and Jacobson, Michael. *Food for People, Not for Profit: A Sourcebook on the Food Crisis.* N.Y.: Ballantine, 1975. Action handbook on a wide variety of food issues.

McConnell, Grant. *The Decline of Agrarian Democracy.* Berkeley: Univ. of California Press, 1959.

Madden, J. Patrick. *Economies of Size in Farming: Theory, Analytical Procedures, and a Review of Selected Studies.* AER-107. Washington, D.C.: ERS, USDA, February 1967. Concludes that the family farm can achieve unit costs as low as, if not lower than, giant corporations.

Marshall, Ray, and Thompson, Allen. *Status and Prospects of Small Farmers in the South.* Atlanta: Southern Regional Council, Inc., 1976. Good summary of the plight and prospects of family farms.

Merrill, Richard, ed. *Radical Agriculture.* N.Y.: Harper Colophon Books, 1976. A presentation of alternatives to prevailing practices in American agriculture.

Morgan, Dan. *Merchants of Grain.* N.Y.: Viking, 1979. Offers an illuminating account of how the grain trade works and how it affects the world's eating habits.

National Farm Institute, ed. *Corporate Farming and the Family Farm.* Ames: Iowa State Univ. Press, 1970.

National Rural Center. *Toward a Federal Small Farms Policy, Phase I: Barriers to Increasing On-Farm Income.* Washington D.C., 1978. Focus on how the "small farm family" should be defined, and what are the key problems hindering these farm families from increasing their earnings.

Nikolitich, Radoje. *Family-Size Farms in U.S. Agriculture.* ERS-499. Washington, D.C.: ERS, USDA, February 1972. Focuses on changes in the relative position of family farms in American agriculture.

Orden, David, and Smith, Dennis K. *Small Farm Programs: Implications From a Study in Virginia.* Research Division Bulletin 135. Blacksburg, Va.: VPI, 1978. Focus on the low income of small farmers and how the incomes can be increased.

Paarlberg, Don. *Farm and Food Policy: Issues of the 1980s.* Lincoln: Univ. of Nebraska 1980. Considered are commodity programs, price control, decline of the family farm, and the rise of agribusiness, the role of the USDA, and international trade policy.

Pavlick, Anthony L. *Towards Solving the Low-Income Problem of Small Farmers in the Appalachian Area.* Bulletin 499T. West Virginia Agricultural Experiment Station, June 1964. Assesses the economic position of Appalachian farmers and offers reform suggestions.

Perelman, Michael. *Farming for Profit in a Hungry World: Capital and the Crisis in Agriculture.* Montclair, N.J.: Allanheld, Osmun & Co., 1977. Examines the costs associated with modern American agricultural methods.

Pierce, Dean. "The Low-Income Farmer: A Reassessment." *Social Welfare in Appalachia* 3 (1971): 7-10. Concludes that the larger community has misunderstood the historical development of the Appalachian subsistence farmer and what might be needed to aid him.

Policy Studies Journal 6:4 (1977-78). Articles on the political economy of United States agriculture.

Proctor, Roy E., and White, T. Kelley. "Agriculture: A Reassessment." In *The Southern Appalachian Region: A Survey.* Ed. Thomas Ford. Lexington: Univ. of Kentucky Press, 1962. General survey of Appalachian agriculture.

Raup, Philip M. "Corporate Farming in the United States." *Journal of Economic History* 33 (March 1973): 274-90. Provides a survey of the history of corporate farming in America and suggests a number of interesting conclusions about its rise.

Ray, Victor K. *The Corporate Invasion of American Agriculture.* Denver: National Farmers Union, 1968. Concise summary of the dimensions and effects of the invasion of American agriculture by corporate nonfarm interests.

Ritchie, Mark. *The Loss of Our Family Farms.* San Francisco: By the Author, 1979. Identifies the causes for the decline of the small farm.

Robbins, William. *The American Food Scandal: Why You Can't Eat Well on What You Earn.* N.Y.: Morrow, 1974. Critical of the USDA and agribusiness.

Shover, John. *First Majority—Last Minority: The Transforming of Rural Life in America.* Dekalb: Northern Illinois Univ. Press, 1976. Small farmers are today's last "minority," and with their departure rural life has been transformed.

Smith, Charles L. *A Bibliography on the 160-Acre Anti-Monopoly Water Law.* San Francisco: Center for Rural Studies, 1974. Useful bibliography on this important issue.

Southern Land Economic Research Committee. *Farmland Tenure and Farmland Use in the Tennessee Valley.* Pub. No. 9, n.p., n.d. Includes Robert Boxley and Joel Bass, "Land Tenure Institutions in the Tennessee Valley."

Stavrianos, L. S. *The Promise of the Coming Dark Age.* San Francisco: Freeman, 1976. Makes a strong case for the viability of the family farm in the future.

Stewart, Fred J.; Hall, Harry H.; and Smith, Eldon D. "Potential for Increased Income on Small Farms in Appalachian Kentucky." *American Journal of Agricultural Economics* 61 (February 1979): 77-82. Describes the characteristics of low-income farms and offers several suggestions for increasing income.

U.S. Congress. House of Representatives. Subcommittee on Family and Rural Development. Special Study of the Committee on Agriculture, Hearings on

Obstacles to Strengthening Family Farm Systems. 95th Congress, 1st session, 1977. Important collection of statements on why the small farm is disappearing.

U.S. Congress. Senate. Committee on the Judiciary. Hearings on *Priorities in Agricultural Research of the USDA.* 95th Congress, 1st session. Washington, D.C.: USGPO, 1977. Hearings centered around complaints that USDA research is socially and environmentally irresponsible.

U.S. Congress. Senate. Select Committee on Small Business and the Committee on Interior and Insular Affairs. *Will the Family Farm Survive in America?* 4 vols., 3 parts. Hearings, 94th Congress, 2nd session. Washington, D.C.: USGPO, 1975-76.

USDA. *Living on a Few Acres. The 1978 Yearbook of Agriculture.* Washington, D.C., 1979. Focus on the small farmer.

USDA, ERS. *Employment, Unemployed and Low Incomes in Appalachia.* Washington, D.C., 1965. Includes a discussion of the situation of the small farmer.

USDA, ESCS, *Small-Farm Issues: Proceedings of the ESCS Small-Farm Workshop,* May 1978. ESCS-60. Washington, D.C., 1979.

———. *Status of the Family Farm.* Washington, D.C.: USDA, September 1978. Examines the trends and the forces affecting the structure of farming.

———. *Status of the Family Farm:* Second Annual Report to the Congress. AER-434. Washington D.C., 1979. Update of 1978 report.

———. *Structure Issues of American Agriculture.* AER-438. Wash. DC, 1979. Sections on farm production, public policies, marketing, and the experiences of others.

U.S. Dept. of Commerce. Bureau of Census. *1978 Census of Agriculture.* Washington, D.C., USGPO, 1981. Most recent agricultural census.

U.S. General Accounting Office. *Some Problems Impeding Economic Improvement of Small Farm Operations: What the Department of Agriculture Could Do.* Washington, D.C.: USGPO, August 1975. Recommendations for improving extension services and the land-grant college complex so as to benefit small farms.

Wellford, Harrison. *Sowing the Wind: Ralph Nader's Study Group Report on Food Safety and the Chemical Harvest.* N.Y.: Grossman, 1972. Looks at policies related to pesticides, meat and poultry inspection. Highly critical of the USDA.

Zwerdling, Daniel. "The Food Monsters: How They Gobble Up Each Other—and Us." *Progressive* (March 1980), 16-27. Interesting look at the food industry and its effect on farmers.

5. GOVERNMENT LANDOWNERSHIP, ESPECIALLY NATIONAL FORESTS

Works concerned with the general history of public land development are included in part 1.

Advisory Commission on Intergovernmental Relations. *The Adequacy of Federal Compensation of Local Governments for Tax Exempt Federal Lands.* Washington, DC, 1978. Includes a very useful bibliography.

Appendix Three

Barney, Daniel, R. *Last Stand: Ralph Nader's Study Group Report on the U.S. Forest Service.* N.Y.: Grossman, 1974. Critical survey of USFS history and current policies.

Blanton, Bill. "The National Recreation Area: What Will It Mean?" *The Plow* (1 June 1978): 15-18, 24-26. Also in F, pp. 181-85. Critically examines the rationale of the USFS's plan to build a scenic highway in the Mount Rogers National Forest. For more information on this, see 1978 and 1979 issues of *The Plow*.

Cameron, Jenks. *The Development of Governmental Forest Control in the United States.* Baltimore: JHP, 1928; repr. N.Y.: DaCapo, 1972. Major study of national forest policy since the colonial period.

Citizen Handbook on Forest Management: Ozark National Forests. Fayetteville and Eureka Springs, Ark.: The Sierra Club, Newton County Wildlife Association, and The Ozark Institute, 1977. Educational overview of forest management for use by citizen groups.

Clawson, Marion, *Decision Making in Timber Production, Harvest, and Marketing.* Washington, D.C.: RFF, 1977. Attempt to clarify some of the problems and opportunities of timber growing.

———. *The Economics of National Forest Management.* Published for RFF. Baltimore: JHP, 1976. Considers costs and outputs of national forest management.

———. *The Federal Lands Since 1956: Recent Trends in Use and Management.* Published for RFF. Baltimore: JHP, 1967. Update of data and analysis in Clawson and Held's *The Federal Lands* (1957).

———. *Forests for Whom and for What?* Published for RFF. Baltimore: JHP, 1975. Analysis of the criteria that should be used to shape USFS policy.

———. *Forest Policy for the Future: Conflict, Compromise, Consensus.* Papers and discussion from a Forum on Forest Policy for the Future, 8-9 May 1974, Washington, D.C. Washington, D.C.: RFF, 1974. Attempt to formulate a forest policy for the nation.

Clawson, Marion, and Held, R. Burnell. *The Federal Lands: Their Use and Management.* Published for RFF. Baltimore: JHP, 1957. Examines the management problems of the federal lands and the governmental procedures by which such management problems are decided. Describes the economic impact of major policies and reviews the operation of pertinent laws and regulations. Makes the case for a federal land corporation.

Clepper, Henry. *Professional Forestry in the United States.* Published for RFF. Baltimore: JHP, 1971. Useful topical history.

Conservation Foundation. *National Parks for the Future: An Appraisal of the National Parks as They Begin Their Second Century in a Changing America.* Washington, D.C., 1972. Collection of essays that examine national park issues and make recommendations on how to ensure that the parks will be used for the benefit and use of the people.

Cooper, Diana S., and Worrell, Albert C. "Forest Land as Taxable Property." *Journal of Forestry* 69 (July 1971): 401-06. Concludes that forestland situations differ so much that different approaches are necessary.

Dana, Samuel T. *Forest and Range Policy: Its Development in the United States.* N.Y.: McGraw-Hill, 1956. Well-researched history of American forestry and forest legislation from the colonial period to the 1950s. Includes a useful chronological summary of important events.

Davis, Richard C. *North American Forest History: A Guide to Archives and Manuscripts in the United States and Canada.* Santa Barbara, Cal.: A.B.C.-Clio Press, 1977. Most comprehensive description of unpublished materials dealing with the USFS.

Duerr, William A. *The Economic Problems of Forestry in the Appalachian Region.* Harvard Economic Studies, No. 84. Cambridge: Harvard Univ. Press, 1949. A regional approach that attempts to view the forestry problems of each locality against the background of the economy of the whole region.

———, ed. *Timber! Problems, Prospects, Policies.* Ames: Iowa State Univ. Press, 1973. Examines key timber issues.

Fahl, Ronald J. *North American Forest and Conservation History: A Bibliography.* Santa Barbara, Cal.: A.B.C.–Clio Press, 1977. Comprehensive bibliography on USFS history that was very helpful in compiling this section of the bibliography.

Finger, Bill; Fowler, Cary; and Hughes, Chip. "Tree Killers on the Rampage." *Southern Exposure* 2 (Fall 1974): 170-77. Describes the hold that large timber companies have on forestland in the South.

Frome, Michael. *The Forest Service.* N.Y.: Praeger, 1971. Informal and slightly critical discussion of USFS issues and history.

Hagenstein, Perry R. *Forestry and the Appalachian Regional Commission.* A report to the ARC. Washington, D.C., May 1974. Makes recommendations for ARC policy.

Hahn, Benjamin; Post, J. Douglas; and White, Charles B. *National Forest Resource Management: A Handbook for Public Input and Review.* Stanford: Stanford Environmental Library, 1978. Intended as a practical guide for citizen participation in USFS land-mangement planning.

Hargraves, L. A., Jr., and Jones, Richard W. *Forest Property Taxation.* Report No. 29. Georgia Forest Research Council, May 1972. Examines various historical theories and assumptions of the property tax, discusses the general nature of the property tax today as it relates to forestry, and offers a model for rural property taxation.

Horowitz, Eleanor C. J., ed. *Clearcutting: A View from the Top.* Washington, D.C.: Acropolis Books, 1974. Essays on the effects of clearcutting. Bibliography.

Ise, John. *The United States Forest Policy.* New Haven: Yale Univ. Press, 1920; repr. N.Y.: Arno, 1972. Pioneering study of early national forest policy.

Kahn, Si. *The Forest Service and Appalachia.* N.Y.: John Hays Whitney Foundation, 1974. See also "The Government's Private Forests." *Southern Exposure* 2 (Fall 1974): 132-44. Survey of USFS ownership and its effects on the local tax base.

Kaufman, Herbert. *The Forest Ranger: A Study in Administrative Behavior.* Published for RFF. Baltimore: JHP, 1960. Insight into how the decentralized USFS maintains control over its field personnel.

Kurtz, William B., and Lapping, Mark B. *The Small Forest Land Owner: Ownership and Characteristics.* Exchange Bibliography 1113. Monticello, Ill.: Council of

Planning Librarians, September 1976. About 59 percent of the nation's commercial-grade forestland is held by the nonindustrial private landowner.

Lowry, Stanley T. "The Appalachian Regional Development Act of 1965 and the Timber Development Organization Dilemma." *Journal of Forestry* 64 (December 1966): 785-91. Isolates some of the frustrating problems of forestry's role in economic development.

McDonald Associates, Inc. *Evaluation of Timber Development Organizations.* Appalachian Research Report No. 1. Washington D.C.: ARC, 1 November 1966. Examines ownership, condition, and use of timber within Appalachia and makes recommendations for the establishment of a timber development program to be carried out by private, nonprofit corporations.

McWhinney, William. *The National Forest: Its Organization and Its Professionals.* Los Angeles: UCLA Graduate School of Business Administration, 1970. Includes a field survey of nearly 1200 professionals and 600 technicians.

Munns, Edward N. *A Selected Bibliography of North American Forestry.* 2 vols. MP-364. Washington, D.C.: USGPO, 1940. Good source for technical and periodical literature.

Neiderheiser, Clodaugh M. *Forest History Sources of the United States and Canada: A Compilation of the Manuscript Sources of Forestry, Forest Industry, and Conservation History.* St. Paul, Minn.: Forest History Foundation, 1956. Cites 972 groups of manuscripts.

Nienaber, Jeanne, and Wildavsky, Aaron. *The Budgeting and Evaluation of Federal Recreation Programs or Money Doesn't Grow on Trees.* N.Y.: Basic Books, 1973. Focuses on the Forest Service, National Park Service, and the Bureau of Outdoor Recreation.

Ogden, Gerald. *The United States Forest Service: A Historical Bibliography, 1876-1972.* Davis: Agricultural History Center, Univ. of California, June 1976. Lists over 7,000 items.

Osborn, William C. *The Paper Plantation: Nader Study Group Report on the Pulp and Paper Industry in Maine.* NY: Grossman, 1974. Reports that seven absentee corporations own 32 percent of Maine's 20 million acres.

O'Toole, Randal Lee. *The Citizen's Guide to Forestry and the Environment.* Portland, Ore.: Cascade Holistic Economic Consultants, 1977. Provides tools for citizens to use when becoming involved in forest-management issues.

Pinchot, Gifford. *Breaking New Ground.* N.Y.: Harcourt, 1947; repr. Seattle: Univ. of Washington Press, 1972. Autobiography of this central figure in USFS history.

Public Land Law Review Commission. *One Third of the Nation's Land: A Report to the President and to the Congress.* Washington, D.C.: USGPO, June 1970. Offers recommendations for policy guidelines for the retention and management or disposition of federal lands.

Report of the President's Advisory Panel on Timber and the Environment. Washington, D.C.: USGPO, April 1973. Report of a special ad hoc panel, containing recommendations for increasing timber supply while maintaining the forest environment for other uses.

Risser, James. "The U.S. Forest Service: Smokey's Strip Miners." *Washington Monthly* 3 (December 1971): 16-26. Examines the link between the USFS and the National Forest Productions Association, the industry lobby, especially in relation to the clearcutting issue.

Robinson, Glen O. *The Forest Service: A Study in Public Land Management.* Published for RFF. Baltimore: JHP, 1975. Useful history, which includes a discussion of the major land-management problems facing the USFS.

Rose, John Kerr. *Survey of National Policies on Federal Land Ownership.* Senate Report, 85th Congress, 1st session. Doc. No. 338. Washington, DC: USGPO, 1957. Surveys historic policies on federal lands, as well as studies proposals and bills to divest certain public lands to the states and private economy.

Schenck, Carl A. *The Birth of Forestry in America: Biltmore Forest School 1898-1913.* Ed. Ovid Butler. Santa Cruz, Cal: Forest History Society and the Appalachian Consortium, 1974. Story of America's first forestry school.

Schiff, Ashley L. *Fire and Water: Scientific Heresy in the Forest Service.* Cambridge: Harvard Univ. Press, 1962. Examines controversies within the USFS over controlled burning as a forest management practice and over the relation of forests to flood control.

Shands, William E., and Healy, Robert G. *The Lands Nobody Wanted: Policy for National Forests in the Eastern United States.* Washington D.C.: Conservation Foundation, 1977. Calls attention to the distinctive qualities of the Eastern national forests and the various demands placed upon them and suggests long-range management policies sensitive to these lands.

Shaw, Elmer W. "Pro and Con Analysis of Clearcutting as a Forestry Practice in the United States." Washington, D.C.: Congressional Research Service, Library of Congress, 15 July 1970. Includes a helpful bibliography.

Shepherd, Jack. *The Forest Killers: The Destruction of the American Wilderness.* N.Y.: Weybright, 1975. Muckraking account of how private interests, especially lumber companies, have been allowed to exploit the national forests.

Smith, Darrell H. *The Forest Service: Its History, Activities and Organization.* Washington, D.C.: Brookings, 1930; repr. N.Y.: AMS Press, 1973. General history of the USFS.

Spurr, Stephen H. *American Forest Policy in Development.* Seattle: Univ. of Washington Press, 1976. Lectures on how USFS policy should evolve.

Steen, Harold K. *The U.S. Forest Service: A History.* Seattle: Univ of Washington Press, 1976. One of the more recent and informative histories of the USFS. Includes a bibliographical essay.

Stoddard, Charles H. *The Small Private Forest in the United States.* Published for RFF. Baltimore: JHP, 1961. Examines the problem of the unproductive, small private forests and makes suggestions for increasing their productivity.

U.S. Congress. Senate. Committee on Interior and Insular Affairs, Subcommittee on Public Lands. *Clearcutting on Federal Timberlands: A Report.* Washington, DC: USGPO, 1972. Contains valuable information on the clearcutting issue.

——— *"Clearcutting" Practices on National Timberlands.* 3 Parts. Hearings, 92nd Congress, 1st session. Washington, D.C.: USGPO, 1971. Useful summary of the opposing views in the clearcutting debate.

USDA. *Message from the President of the United States Transmitting a Report of the Secretary of Agriculture in Relation to the Forests, Rivers, and Mountains of the Southern Appalachian Region.* Washington, DC: USGPO, 1902. Makes the case for the government purchase of forest lands in Southern Appalachia.

USDA, USFS. *Guide for Managing the National Forests in the Appalachians.* 2nd ed. Washington, D.C., 1973. Useful overview of national forests in Appalachia.

———. *The Principal Laws Relating to Forest Service Activities.* rev. ed. Agricultural Handbook No. 453. Washington, D.C., September 1978. Useful compendium of USFS legislation.

U.S. General Accounting Office. *The Federal Drive to Acquire Private Lands Should Be Reassessed.* Washington, D.C.: USGPO, December 1979. Highly critical of government acquisition policies.

Widner, Ralph K., ed. *Forests and Forestry in the American States: A Reference Anthology.* Washington, D.C.: National Association of State Foresters, 1968. Contains many historical articles on forest conservation, state forestry, and state forestry organizations.

Winters, Robert K., ed. *Fifty Years of Forestry in the U.S.A.* Washington, D.C.: Society of American Foresters, 1950. Nineteen essays on progress in American forestry.

Williams, Ellis T. "Emerging Patterns of Forest Tax Legislation." *Agricultural Finance Review* 32 (August 1971): 15-21. Summarizes the legislative trends in the property taxation of forestland and tenure.

Wood, Nancy. *Clearcut: The Deforestation of America.* San Francisco: Sierra Club, 1971. Critical analysis of the effects of clearcutting.

Yoho, James G. "The Responsibility of Forestry in Depressed Areas." *Journal of Forestry* 63 (July 1965): 508-12. Outlines a "more rational philosophy" to guide efforts to use forest resources as a "vehicle of economic uplift" in economically depressed areas.

6. FEDERAL AND PRIVATE DAM BUILDERS

American Rivers Conservation Council. *How to Save Your River: A Citizen's Guide to Water Projects.* Washington, D.C., n.d. Also in F, pp. 201-05. Brief guide about how citizens can organize to halt a destructive water project proposed by the U.S. Army Corps of Engineers.

American Rivers Conservation Council and the Environmental Policy Center. "95 Theses." Washington, D.C., May 1974. Call for reform in the U.S. Army Corps of Engineers and the Bureau of Reclamation.

Berkman, Richard L., and Viscusi, W. Kip. *Damming the West: Ralph Nader's Study Group Report on the Bureau of Reclamation.* N.Y.: Grossman, 1973. Highly critical analysis of the destructive practices of this agency.

Blackwelder, Brent. *Benefit Claims of the Water Development Agencies: The Need for Continuing Reform.* Washington, D.C.: Environmental Policy Institute, 1976. Addresses several of the major areas of controversy surrounding the water resources projects of the U.S. Army Corps of Engineers.

———. *Citizens' Guide to the New Carter Water Policy.* Washington, D.C.: Environmental Policy Center, 1978. Critique of the national water policy announced in June 1978.

Blackwelder, Brent, and Carlson, Peter. *An Analysis of the Stonewall Jackson Lake Project of the U.S. Army Corps of Engineers.* Washington, D.C.: Environmental

Policy Center, 1979. Highly critical evaluation of Corps's actions in this West Virginia water project.

Blanton, Bill. "Not by a Dam Site: Brumley Gap, Virginia—How One Community Fought Back." *Southern Exposure* 7 (Winter 1979): 98-106. Good summary of citizen resistance to Appalachian Power Company's attempt to build a pumped-storage facility at Brumley Gap, Virginia

Branscome, Jim. *The Federal Government in Appalachia.* N.Y.: Field Foundation, 1977. TVA section discusses how the agency has removed over 125,000 people from their land to build dams.

Delaware Valley Conservation Association. *In Defense of Rivers.* Stillwater, N.Y., n.d. Up-to-date guide for fighting destructive water projects.

Douglas, William O. "The Corps of Engineers: The Public Be Damned." *Playboy* (July 1969): 143, 182-88. Critical discussion of the Corps by a former Supreme Court justice.

Drew, Elizabeth. "Dam Outrage: The Story of the Army Engineers." *Atlantic* (April 1970): 51-62. Describes how the Corps wins over $1 billion a year from Congress to build projects that frequently serve only the narrowest interests and too often inflict the wrong kinds of change on the environment.

Eckstein, Otto. *Water-Resources Development: The Economics of Project Evaluation.* Cambridge: Harvard Univ. Press, 1961. Analyzes the procedures for measuring benefits and costs employed by the Bureau of Reclamation and the Corps of Engineers. Assumes that judgments about changes in the distribution of income and about political and social objectives must be left to Congress and that the national interest is best served by benefit-cost analyses which reveal each project's impact on the total real national income to be enjoyed by each county.

Ferejohn, John A. *Pork Barrel Politics: Rivers and Harbors Legislation. 1947-1968.* Stanford: Stanford Univ. Press, 1974. Focuses on Congress's role in the Corps of Engineers appropriations process and the distribution of its projects.

Field, Donald R.; Barron, James C.; and Long, Burl F., eds. *Water and Community Development: Social and Economic Perspectives.* Ann Arbor, Mich. Ann Arbor Science Pubs., 1974. Explores the role of water resources in contributing to community development.

Fitzsimmons, Stephen J. *Man and Water: A Social Report.* An Abt Associates Study in Applied Social Research. Boulder, Col.: Westview, 1977. Examines the relationship between man's social-psychological needs and his water-resources development.

Gaillard, Frye. "Fear of a Final Solution." *Race Relations Reporter 5* (March 1974): 13-14. Concludes that the methods and motives of the Corps of Engineers and others raise the prospect of the forced relocation of mountain people to allow undisturbed access to natural resources.

Garrison, Charles B. "A Case Study of the Local Economic Impact of Reservoir Recreation." *Journal of Leisure Research* 6 (Winter 1974): 7-19. Challenges the belief that Corps of Engineers' dams and reservoirs add to regional productivity.

Giefer, Gerlad J. *Sources of Information in Water Resources: An Annotated Guide to Printed Materials.* Port Washington, N.Y.: Water Information Center, 1976.

Annotates over 1,100 titles found useful for reference purposes in the water-resources field.

Haveman, Robert H. *The Economic Performance of Public Investments: An Ex Post Evaluation of Water Resources Investments.* Published for RFF. Baltimore: JHP, 1972. Presents a general framework for evaluating the direct benefits and costs of a water project after it is completed.

———. *Water Resources Investment and the Public Interest.* Nashville: Vanderbilt Univ. Press, 1965. Analysis of Corps of Engineers expenditures in ten Southern states.

Haveman, Robert H., and Hamrin, Robert D., eds. *The Political Economy of Federal Policy.* N.Y.: Harper, 1973. Includes several articles critical of the Corps of Engineers.

Haveman, Robert H., and Krutilla, John. *Unemployment, Idle Capacity, and the Evaluation of Public Expenditures: National and Regional Analyses.* Published for RFF. Baltimore: JHP, 1968. Examines the adequacy of benefit-cost analysis for evaluating water projects during periods of less than full employment in both the national and regional economies. Presents a methodology for tracing the incidence of water development resource demands generated by regions and sectors of the national economy.

Holmes, B. H. *History of Federal Water Resources Programs and Policies, 1961-1970.* MP-1379. Washington, D.C.: USDA, 1979.

Holt, William S. *The Office of the Chief of Engineers of the Army: Its Non-Military History, Activities and Organization.* Baltimore: JHP, 1923; repr. N.Y.: AMS Press, 1974. Early history of the Corps.

Howe, Charles W. *Benefit-Cost Analysis for Water System Planning.* Water Resources Monograph No. 2. Washington, D.C.: American Geophysical Union, 1971. Sets forth some of the basic elements of a broad benefit-cost approach to water-resources planning.

Jacobstein, J. Myron, and Mersky, Roy M. *Water Law Bibliography, 1847-1965: Source Book on U.S. Water and Irrigation Studies—Legal, Economic and Political.* Silver Spring, Md.: Jefferson Law Book Co., 1966. Updated through 3 supplements—1966-67, 1968-73, and 1974-77.

Johnson, Leland R. "A History of the Operations of the Corps of Engineers, United States Army, in the Cumberland and Tennessee River Valleys." Ph.D. Dissertation, Vanderbilt Univ., 1972. Includes useful information on the extent of Corps activities in this section of Appalachia.

"Kentucky Rivers Coalition." *MLW* (November 1976): 26-27. Describes the activities of this important citizen group.

Laycock, George. *The Diligent Destroyers.* Garden City, N.Y.: Doubleday, 1970. Includes a section on the Corps of Engineers.

Leopold, Luna B., and Maddock, Thomas, Jr. *The Flood Control Controversy: Big Dams, Little Dams and Land Management.* Sponsored by the Conservation Foundation. N.Y.: Ronald, 1954. Examines flood-control programs and controversies involving the Corps of Engineers, USFS, and USDA.

Maass, Arthur. *Muddy Waters: The Army Engineers and the Nation's Rivers.* Cambridge: Harvard Univ. Press, 1951; repr. N.Y.: DaCapo, 1974. A history of the Corps.

Marine, Gene. *America the Raped: The Engineering Mentality and the Devastation of a Continent.* N.Y.: Simon & Schuster, 1969. Includes a discussion of several Corps of Engineers projects.

Massey, David. "Over a Barrel: Southern Waterways and the Army Corps of Engineers." *Southern Exposure* 8 (Spring 1980): 92–100. Excellent critique of Corps projects in the South during the 1970s.

Matthiessen, Peter. "How to Kill a Valley." *New York Review of Books* (7 February 1980): 31-36. Good critique of TVA's Tellico Dam project.

Mazmanian, Daniel, and Nienaber, Jeanne. *Can Organizations Change? Environmental Protection, Citizen Participation and the Corps of Engineers.* Washington, D.C.: Brookings, 1979. Examines how the Corps responded to pressures for greater public involvement and environmental awareness in its decision-making structure.

Morgan, Arthur E. *Dams and Other Disasters: A Century of the Army Corps of Engineers in Civil Works.* Boston: Porter Sargent, 1971. Very critical account of the Corps and its abuse of the environment.

Napier, Ted L. *An Analysis of the Social Impact of Water Resource Development and Subsequent Forced Relocation of Population Upon Rural Community Groups: An Attitudinal Study.* ESS513. Columbus: Dept. of Agricultural Economics & Rural Sociology, Ohio State Univ., 1974. Examines social impacts within several communities in Ohio and West Virginia that had recently experienced extensive water-development activity.

Parfit, Michael. "The Army Corps of Engineers: Flooding America in Order to Save It." *New Times* (12 November 1976): 25+. Concludes that the Corps "resembles a giant bulldozer out of control, burying villages, disfiguring the landscape and possibly bringing closer the very floods it is meant to prevent."

Reuss, Henry S. "Needed: An About-Face for the Army Corps of Engineers." *Reader's Digest* (November 1971): 129-32. A congressman criticizes the Corps.

Rosenbaum, Walter A. *The Politics of Environmental Concern.* 2nd ed. N.Y.: Praeger, 1977. Includes a chapter on the politics of water resources that emphasizes the activities of the Corps of Engineers.

Saving Rivers. Summary of workshops of the 1978 Conference on Rivers, Dams, and National Water Policy, Washington, D.C. Washington, D.C.: Environmental Policy Center, 1978. Useful collection that examines relevant legislation and explores ways to oppose destructive water projects.

Schoenbaum, Thomas J. *The New River Controversy.* Winston-Salem, N.C.: John F. Blair, 1979. Examines the defeat of Appalachian Power Company's effort to build a pumped-storage facility along the New River in Virginia and North Carolina.

Tennessee-Tombigbee Background Report. Washington, D.C.: Coalition for Water Project Review, March 1979. Summary of the issues involved in this highly controversial water project.

Whisnant, David. *Modernizing the Mountaineer: People, Power, and Planning in Appalachia.* N.Y.: Burt Franklin, 1980. Includes a discussion on TVA's dam-building philosophy.

Wilkins, Bryan. "Tennessee-Tombigbee: New American Flush." *Mountain Review* 3 (September 1977): 1-5. Critical account of this Corps of Engineers project.

7. LAND-REFORM PROPOSALS AND STRATEGIES

This is a brief section for several reasons. First, many of the works listed in the above sections include reform proposals and, with just a few exceptions, these works are not repeated here. Second, land reform has not been a serious issue in America for many years and the literature related to this issue is woefully inadequate. Finally, except for a few bibliographies and general surveys, works on land-reform efforts in other nations are not inlcuded.

Barnes, Peter, ed. *The People's Land: A Reader on Land Reform in the United States.* Emmaus, Pa.: Rodale, 1975. Significant collection of articles on land-reform issues.

Barnes, Peter and Casalino, Larry. *Who Owns the Land? A Primer on Land Reform in the USA.* Berkeley, Cal.: Center for Rural Studies, 1972. Focus on various land-reform alternatives.

Belden, Joe; Edwards, Gibby; Guyer, Cynthia; and Webb, Lee, eds. *New Directions in Farm, Land and Food Policies: A Time for State or Local Action.* Washington, D.C.: Agricultural Project, Conference on Alternative State and Local Policies, 1978. Summary of reform efforts at the state and local level.

Bruchey, Stuart. "The Twice 'Forgotten' Man: Henry George." *American Journal of Economics and Sociology* 31 (April 1972): 113-38. Examines the relation of George's thoughts on the land issue to his times and our own.

Center for Science in the Public Interest. *From the Ground Up: Building a Grass Roots Food Policy.* Washington, D.C., 1976. Action handbook for people who want to reform food policies at the city, county, and state levels.

Chambers, H. Harold, Jr. *Citizen Participation in the Battle Against Strip Mining Under the West Virginia Surface Mining and Reclamation Act of 1971.* Charleston, W.Va.: Appalachian Research and Defense Fund, 1975. Handbook for fighting strip-mining.

Citizens' Energy Project. "Land Use Options for Citizens Impacted by Appalachian Coal Development." Washington, D.C., n.d. Considers several legal options available to citizens and local and state governments to respond to some of the adverse land-use impacts brought about by coal development. Focuses on West Virginia, Kentucky, and Tennessee.

Constant, Florence. *Land Reform: A Bibliography.* Cambridge, Mass.: Center for Community Economic Development, 1972. Nine-page bibliography.

Corty, Floyd L. "Are We Headed for Land Reform in the United States?" *Land Economics* 38 (August 1962): 270-73. Concludes that the possibilities of land reform here are remote.

Curie, Robert F. "Columbia Gas Ordered to Pay Higher Taxes." *MLW* (April 1979): 3-6. Also in F, pp. 222-25. Describes how a local citizens' group won a major victory against a large corporate landholder.

Dorner, Peter. *Land Reform and Economic Development.* Baltimore: Penguin, 1972. Treats land reform as an integral part of the strategy and policy of economic development. Useful bibliography.

Faux, Geoffrey. "Reclaiming America." *Working Papers* 1 (Summer 1973): 31-42. Frequently cited article on land reform in America.
Fisher, Steve, and Foster, Jim. "Models for Furthering Revolutionary Praxis in Appalachia." *Appalachian Journal* 6 (Spring 1979): 171-96. Includes a section on resistance movements formed around the land issue.
Food and Agriculture Organization of the United Nations, Documentation Centre. *Land Reform: Annotated Bibliography of FAO Publications and Documents (1945-April 1970).* N.Y.: Unipub, Inc., 1971. Valuable reference for non-American sources.
──. *Land Reform, Land Settlement and Cooperatives.* Issued semi annually. No. 1 (1968). A medium for dissemination of information and views on land reform and related subjects.
Fritsch, Albert J. *The Contrasumers: A Citizen's Guide to Resource Conservation.* N.Y.: Praeger, 1974. Examines a set of conservation strategies.
George, Henry. *Progress and Poverty.* NY: Robert Schalkenbach Foundation, 1965 (orig. 1887). Major work of this advocate for the nationalization of the value of land.
Guitar, Mary Anne. *Property Power: How to Keep the Bulldozer, the Power Line and the Highwaymen From Your Door.* Garden City, N.Y.: Doubleday, 1972. Helpful handbook for organizing around land issues.
Hardt, Jerry (with Diane Sternberg, Linda Kravitz, and Dave Kirkpatrick). *The Feasibility and Design of a Central Appalachian Land Bank.* Prepared for ARC. Berea, Ky: Human/Economic Appalachian Development Corporation, 1979. Makes the case for a land bank and sets forth a strategy for establishing one in Central Appalachia.
"Housing Alternative." *Elements* (February 1977): 8-10. Argues for the public acquisition of land for houses.
Illinois South Project. *A Handbook on Coal Leasing and Landowners Organizations.* Carterville, Ill, 1976. A handbook for local citizens. Makes recommendations for limiting the coal mining of agricultural land.
International Independence Institute. *The Community Land Trust: A Guide to a New Model for Land Tenure in America.* Cambridge, Mass.: Center for Community Economic Development, 1972. Summarizes the key issues in the land-trust movement.
Jacoby, Erich H. In collaboration with Charolotte F. Jacoby. *Man and Land: The Essential Revolution.* N.Y.: Knopf, 1971. Describes current conditions of the man-land relationship and makes the case for land reform. Includes a useful bibliography.
Kahn, Si. *How People Get Power: Organizing Oppressed Communities for Action.* N.Y.: McGraw-Hill, 1970. Focus on rural communities.
King, Russell. *Land Reform: A World Survey.* London: G. Bell & Sons, 1977. Discussion of land reform in Latin America, Asia, Africa, and the Middle East.
Landau, Norman J., and Rheingold, Paul D. *The Environmental Law Handbook.* N.Y.: Friends of the Earth/Ballantine Books, 1971. Legal action simplified for citizen groups.
"Land Reform." *Citizens' Appalachia* No. 12 (January 1979). Summary of land reform issues in Appalachia.

Little, Charles E. *Stewardship.* N.Y.: Open Space Institute, 1965. Account of how land philanthropy benefits both the donor and his community.
Love, Sam, ed. *Earth Tool Kit: A Field Manual for Citizen Activists.* N.Y.: Pocket Books, 1971. Includes useful suggestions for organizing around land issues.
McHarg, Ian. *Design With Nature.* Garden City, N.Y.: Doubleday, 1969. Classic work whose ideas have been used to form a philosophy of land use where planners allow nature to dictate the design.
Rural America. *Toward a Platform for Rural America.* Washington D. C., 1978. Includes a summary of key land issues and proposes a number of reforms.
Schumacher, E. F. *Small Is Beautiful: Economics as if People Mattered.* N.Y.: Perennial Library, 1973. Well-known statement of the importance of small-scale industrial and agricultural production.
Strong, Ann L. *Land Banking—European Reality, American Prospect.* Baltimore: JHP, 1979. Examines the European example and speculates on the potential for land banking in America.
Tai, Hung-Chao. *Land Reform and Politics: A Comparative Analysis.* Berkeley: Univ. of California Press, 1974. Political analysis of land reform in eight developing countries.
Tuma, Elias H. *Twenty-Six Centuries of Agrarian Reform: A Comparative Analysis.* Berkeley: Univ. of California Press, 1965. Provides an interesting and useful classification of land reform movements.
Warriner, Doreen. *Land Reform in Principle and Practice.* Oxford: Clarendon Press, 1969. A general introduction.
Wellstone, Paul D. *How the Rural Poor Get Power: Narrative of a Grass-Roots Organizer.* Amherst: Univ. of Massachusetts Press, 1978. Examines strategies for organizing that are relevant to bringing people together around land issues.
Whyte, William H. *The Last Landscape.* N.Y.: Doubleday, 1968. Readable history of the evolving techniques being used to save the land.
You Can't Put It Back. A West Virginia Guide to Strip Mine Opposition. Fairmont, W.Va.: Mountain Community Union, 1976. Handbook for opposing strip-mining.

NOTES

CHAPTER 1

1. Erik Eckholm, "The Dispossessed of the Earth: Land Reform and Sustainable Development," *Worldwatch Paper 30* (June 1979), p. 6.
2. Gene Wunderlich and D. Barlow Burke, Jr., "Secrecy and Disclosure of Wealth: Publicity of Real Property Holdings," in *Secrecy and Disclosure of Wealth in Land* (Washington, D.C.: The Farm Foundation in cooperation with the U.S. Department of Agriculture Economics, Statistics, and Cooperatives Service, 1978), p. 1.
3. William R. Lassey, *Planning in Rural Environments* (New York: McGraw-Hill, 1977), p. 87.
4. Peter Barnes and Larry Casalino, *Who Owns the Land? A Primer on Land Reform in the United States* (Berkeley, Cal.: Center for Rural Studies, 1972), p. 2.
5. Quoted in Peter Meyer, "Land Rush: A Survey of America's Land," *Harper's Magazine,* January 1979, p. 48.
6. Marion Clawson, *America's Land and Its Uses* (Baltimore, Md.: Johns Hopkins University Press, 1972), pp. 2-3.
7. Meyer, "Land Rush," p. 49.
8. U.S. Department of Agriculture, Economic Research Service, "Corporate Land Holdings: An Inquiry into a Data Source," quoted in Meyer, "Land Rush," p. 49.
9. Geoffrey Faux, "The Future of Rural Policy," in Peter Barnes, ed., *The People's Land: A Reader on Land Reform in the United States* (Emmaus, Pa.: Rodale Press, 1975), p. 191.
10. Steve Fisher and Mary Harnish, "Losing a Bit of Ourselves: The Decline of the Small Farmer" (paper presented at the 1980 meeting of the Appalachian Studies Conference, East Tennessee State University, Johnson City, Tenn.
11. Cited in Meyer, "Land Rush," p. 49.
12. Ibid., p. 50.
13. Peter Barnes, *The Sharing of Land and Resources in America* (Washington, D.C.: a *New Republic* pamphlet, 1973), p. 4.
14. See, for instance, James Krohe, Jr., "Farm Land Bonanza," *The Nation* (17 February 1979), pp. 176–78.
15. Quoted in *Washington Post,* 14 January 1981.
16. Faux, "Future of Rural Policy," pp. 52-53.
17. Cited in Meyer, "Land Rush," p. 49.
18. Bill Finger, Cary Fowler, and Chip Hughes, "Tree Killers on the Rampage," *Southern Exposure* 2 (Fall 1974): 172.

Notes to Pages 4-10

19. See, for instance, Janet Marinelli, "Missing a RARE Opportunity," *Environmental Action* 10 (February 1979): 4-9.

20. Leo McGee and Robert Boone, eds., *The Black Rural Landowner—Endangered Species: Social, Political and Economic Implications* (Westport, Conn.: Greenwood Press, 1979), p. xix.

21. See U.S., Department of Agriculture, *Foreign Ownership of U.S. Agricultural Land*, AER-447 (1980).

22. Clawson, *America's Land*, p. 146.

23. President's Commission on Coal, *The American Coal Miner: A Report on Community and Living Conditions in the Coalfields* (Washington, D.C.: U.S. Government Printing Office, 1980), p. 54.

24. Barnes and Casalino, *Who Owns the Land*, p. 7.

25. U.S. President's Commission on Coal, *Coal Data Book*, (Washington, D.C.: U.S. Government Printing Office, 1980), p. 122.

26. U.S., Congress, Office of Technology Assessment, *The Direct Use of Coal* (Washington, D.C.: U.S. Government Printing Office, 1979), p. 337.

27. Janet M. Smith, David Ostendorf, and Mike Schechtman, *Who's Mining the Farm?* (Herrin, Ill.: Illinois South Project, 1978), p. 10.

28. U.S., *Statutes at Large*, vol. 92, p. 2886.

29. Clawson, *America's Land*, p. 58.

30. (Washington, D.C.: U.S. Government Printing Office, 1976).

31. Ibid., p. 2.

32. Meyer, "Land Rush," p. 47.

33. Barlow Burke, Jr., and Gene Wunderlich, eds., *Secrecy and Disclosure of Wealth in Land* (Washington, D.C.: The Farm Foundation in cooperation with the U.S. Department of Agriculture Economics, Statistics, and Cooperatives Service, 1978), p. iii.

34. Helen Lewis, "Fatalism or the Coal Industry," *Mountain Life and Work* 46 (December 1970): 7.

35. Cited in Cynthia Guyer, "Land Ownership and Inequality in Appalachia," Senior Thesis, Oakes College, 1975, p. 77.

36. Nancy R. Bain, "The Impact of Absentee Ownership in a Rural Area: A Case Study in Southeast Ohio" (unpublished paper, Department of Geography, Ohio University, Athens, Ohio, 1977), p. 70.

37. Ibid., p. 13.

38. Ralph W. Kline, "The Extent and Characteristics of Absentee Landownership in a Five County Region of Southeastern Ohio" (M. A. thesis, Ohio University, Athens, Ohio, 1977), p. 70.

39. Walter Goldschmidt, *As You Sow: Three Studies in the Social Consequences of Agribusiness* (Glencoe, Ill.: Free Press, 1947). A classic study on the impact of large corporate agricultural holdings on rural communities.

40. Agricultural Marketing Project, "Land Ownership Patterns and Community Development," mimeographed, n.d.

41. Harry Caudill, "Appalachia," in Barnes, *People's Land*, p. 33.

42. Ronald D. Eller, "Industrialization and Social Change in Appalachia, 1880-1930: A Look at the Static Image," in Lewis et al., *Colonialism in Modern America: The Appalachian Case* (Boone, N.C.: The Appalachian Consortium Press, 1978), pp. 40.

43. E. E. Hurt, G. T. Tyron, and J. W. Wilts, *What the Coal Commission Found: An Authoritative Summary* (Baltimore, Md.: The Williams and Wilkins Co., 1925), pp. 90–93.

44. Harold M. Watkins, *Coal and Men* (London: George Allen & Unwin, 1934), p. 258.

45. For other summaries of these studies see: David S. Walls, Dwight B. Billings, Mary P. Payne, and Joe F. Childers, Jr., "Coal Land and Mineral Ownership," in *A Baseline*

Assessment of Coal Industry Structure in the ORBES Region (Washington, D.C.: Office of Research and Development, U.S. Environmental Protection Agency, 1979); and John C. Wells, Jr., "Poverty Amidst Riches: Why People Are Poor in Appalachia" (Ph.D. dissertation, Rutgers University, New Brunswick, N.J. 1977), especially pp. 153-98.

46. Richard M. Kirby, "Kentucky Coal: Owners, Taxes, Profits. A Study in Representation without Taxation" (prepared for the Appalachian Volunteers, 1969), p. 1. See also excerpt in *Applachian Lookout* 1 (October 1969): 19-27.

47. James C. Millstone, "East Kentucky Coal Makes Profits for Owners," in *Appalachia in the Sixties*, ed. David Walls and John Stephenson (Lexington, Ky.: University Press of Kentucky, 1972), p. 74.

48. Paul J. Kaufman, "Poor Rich West Virginia," *New Republic* (2 December 1972), p. 12.

49. J. Davitt McAteer, *Coal Mine Health and Safety: The Case of West Virginia* (New York: Praeger, 1973).

50. Tom D. Miller, *Who Owns West Virginia?* (Huntington, W. Va: The Huntington Publishing Co., 1975), p. 1.

51. Ibid., p. 23.

52. John Gaventa, Ellen Ormond, and Bob Thompson, "Coal Taxation and Tennessee Royalists" (Nashville, Tenn.: Vanderbilt Student Health Coalition, 1971), photocopy.

53. Bruce Privratsky and Jane Randolph, "Coal Taxes in Southwest Virginia" (unpublished report for the U.S. Senate Subcommittee on Intergovernmental Relations from the Concerned Citizens for Fair Taxes, 1973).

54. Carol Schommer, "A Critique of Virginia's Mineral Taxation Program" (unpublished study prepared for consideration by Virginia's Department of Taxation and the House of Delegates' Finance Committee from Virginia Citizens for Better Reclamation [VCBR], 1978.

55. Wells, "Poverty Amidst Riches," p. 188.

56. John Morris, "The Potential of Tourism," in *The Southern Appalachian Region: A Survey*, ed. Thomas Ford (Lexington, Ky.: University of Kentucky Press, 1962), p. 137.

57. See, for instance, Edgar Bingham, "The Impact of Recreational Development on Pioneer Life Styles in Southern Appalachia," *Proceedings of the Pioneer America Society* (1973), pp. 59-68; Jim Branscome and Peggy Matthews, "Selling the Mountains," *Southern Exposure* 2 (Fall 1974): 122-29; Jerome E. Dobson, "The Changing Control of Economic Activity in the Gatlinburg, Tennessee, Area, 1930-1973" (Ph.D. dissertation, University of Tennessee, Knoxville, Tennessee, 1975).

58. William Cary, Molly Johnson, Meredith Golden, and Trip Van Noppen, *The Impact of Recreational Development: A Study of Land Ownership, Recreational Development and Local Land Use Planning in the North Carolina Mountain Region* (Durham, N.C.: North Carolina Public Interest Research Group, 1975), p. 1.

59. Si Kahn, "The Government's Private Forests," *Southern Exposure* 2 (Fall 1974): 132.

60. Memo to Reginald Forsyth, Regional Forester, R-8, from Walter W. Rule, Jr., Public Information Officer, entitled "Roadless Area Review and Evaluation II," August 27, 1978, U.S. Forest Service File, North Carolina Division.

61. Kahn, "Private Forests," p. 133.

CHAPTER 2

1. Robert C. Fellmeth, *The Politics of Land* (New York: Grossman Publishers, 1973), p. 14.

2. The index used here was chosen for this study because of its relative simplicity. The

Notes to Pages 18-35

Gini coefficient, another concentration measure, was also computed. The correlation between the Gini index and the index used here is quite high: Pearson's R = .735 at the .001 level of probability.

3. This relationship, e.g., the larger the holding, the more likely the owner will be absentee, is statistically significant: Chi square = 445 at the .0001 level of probability.

4. Statistically, the correlation between the percentage of a county owned by corporations and the percentage owned by absentee holders is significant. Pearson's R = .593 at the .0001 level. In the thirty-seven high coal counties in the survey, the correlation between corporate and absentee ownership rises even further, to .768 at the .0001 level. For mineral rights, the strength of the relationship increases to .967 at the .0001 level.

5. Statistically, there is a significant relationship between the level of corporate ownership of land and the amount of coal reserves in the ground, such that the greater the reserves the greater the percentage of the county corporately owned. (Pearson's R = .368 at the .0015 level of probability.) A stronger correlation is found with coal production, such that the greater the corporate ownership of land, the greater the coal production. (Pearson's R = .463 at the .0001 level of probability.)

6. The Index of Resource Control was developed for this study to deal with the pattern of separated land and mineral ownership. It is admittedly crude, and is affected by the degree of adequate reporting of mineral ownership in the various counties.

7. Analysis of the top fifty owners in the eighty counties does not, of course, give the complete ownership of these corporations in Appalachia; Continental Oil, for instance, owns vast tracts of land not included in the survey area. Similarly, other companies of a given type may happen to own tracts of land in the sample area, which are smaller than those included in this listing. Nevertheless, a look at these top fifty owners provides a cross-section of the types of corporate owners in the region, while not providing the full extent of their holdings.

8. *Dun's Review of Modern Business,* April 1965, p. 40.

9. Quoted in *Charleston* (West Virginia) *Gazette,* 9 January 1980.

10. *Appalachian Land Ownership Study* (Boone, N.C.: Appalachian State University, 1980), 5:36. Hereafter cited as *ALOS.*

11. Ibid, and also newspaper accounts including Nashville *Tennessean,* 24 December 1974, 5 January 1975, 27 April 1977; *Courier Journal;* 30 June 1977.

12. *ALOS,* 3:2-3.

13. *ALOS,* 3:72.

14. *ALOS,* 5:43.

15. Middlesboro *Daily News,* August 1978.

16. *ALOS,* 6:180.

17. *ALOS,* 5:4 and *Courier Journal,* 8 November 1980.

18. Knoxville *News Sentinel,* 26 November 1978.

19. Printed in Louisville *Courier Journal,* 25 March 1980.

20. The *South Magazine,* June 1980, p. 14.

21. Charlotte *Observer,* 21 January 1980.

22. Courier *Journal,* 29 April 1980.

23. Quoted in *Southern Exposure* 4 (Winter 1979): 29.

24. Chessie Systems' holdings in Appalachia are larger outside the survey area. According to the Miller study for the *Huntington Herald Dispatch* in 1974, Chessie owns 517,636 acres in West Virginia alone.

25. Government and private nonprofit owners were originally coded separately. However, owing to the almost insignificant acreage in the private nonprofit category, the two were combined for presentation.

26. John Gaventa, "Property for Prophets," *Southern Exposure* 4 (Summer-Fall, 1973) p. 102.

27. Statistically, there is a significant correlation between the percentage of a county owned by government and private, nonprofit groups and the level of tourism in the county (measured by percentage of services going to tourism and recreation): Pearson's R = .609 at the .0001 level of probability. On the other hand, there is a slightly negative relationship between this category of ownership and the level of coal reserves in a county, such that the greater the coal reserves, the less the extent of public land. (Pearson's R = .293 at the .02 level.)

28. Mrs. John Sutphin, in an interview with Brenda Bowers, seasonal naturalist, Blue Ridge Parkway, 27 August 1914, pp. N16-17, Oral History Transcriptions for Mabry Mill, 1957-1977, in file box labeled "Original Oral History Transcripts," at Appalachian Resources Center, Oteen Maintenance Area, Blue Ridge Parkway, Oteen, N.C.

CHAPTER 3

1. Dick Netzer, *The Economics of the Property Tax* (Washington, D.C.: The Brookings Institution, 1966), p. 16.

2. U.S. Bureau of the Census. 1977 Census of Governments.

3. Valid land-use data was available in only a portion of the counties, as discussed in the methodology appendix. Consequently, the total acres presented in Table 18 are less than the total surface acres in the sample.

4. Section 40-7-15 of Code of Alabama.

5. See *ALOS*, vol. 6. For a more detailed analysis of the mineral taxation as it applies to Wise County, Virginia, see Sandra E. Williams, "An Argument for a Reform in Methods Used to Assess Minerals Not Under Development in Wise County," mimeographed (October 1980).

6. West Virginia Tax Department, Local Government Relations Division, "Valuation of Coal Property for Ad Valorem Taxation" (n.d.), p. 1.

7. West Virginia Tax Department, Local Government Relations Division, "The West Virginia Coal Appraisal/Assessment Program" (n.d.), p. 1.

8. Criticisms include: (a) royalties are often derived from outdated lease and sales information in counties where the coal was acquired in the early 1900s; (b) per-acre value does not include multiple seams where they occur; and (c) county averages of value per acre are based on a single, least-valuable seam. This conservative methodology probably accounts for the fact that not one appraisal has been challenged in court. This argument regarding the conservative bias is made by West Virginians for Fair and Equitable Assessment of Taxes and "Mineral Rights and Property Taxation in West Virginia" by Mohd. Noor Bin Shamsudin and Dale Colyer, Division of Resource Management, College of Agriculture and Forestry, West Virginia University, July 1979.

9. West Virginia Tax Department, "Valuation of Coal Property," p. 5.

10. Donald Colby and David Brooks, "Mineral Resource Valuation for Public Policy," U.S. Bureau of Mines Information Circular 8422, 1969, p. 1.

11. West Virginia Tax Department, "Valuation of Coal Property," p. 2.

12. See Colby and Brooks, "Mineral Resource Valuation."

13. President's Commission on Coal Report, "Staff Findings," 1980, p. 14.

14. In some cases, federal income taxes are subtracted from the income stream. However, since the two states in the study which use a method comparable to this one do not adjust for federal taxes, that has not been done here.

15. Council of State Governments, *Property Tax: A Primer*, p. 11.

16. U.S. Bureau of The Census, 1977 Census of Governments.

17. U.S. Bureau of the Census, 1977 Census of School Finances, p. 2.

18. When median years of education are correlated with the Gini concentration index for the seventy-two rural counties in the sample, Pearson's R = −.418 at the .0003 level of probability, i.e., the greater the concentration of land, the lower the median years of education. Using the concentration index described in Chapter 2, the relationship is less strong (Pearson's R = −.242 at the .0407 level). However, in the tourism counties, where large plots of federal tax-exempt land contribute to a low tax base, the correlation between the concentration index and median level of education rises to −.526 at the .020 level, and to −.447 where the Gini index is used.

CHAPTER 4

1. For a discussion of the failure to develop a mature economy in West Virginia and its roots in the early economic and political developments in the state during its industrialization, see John A. Williams, *West Virginia and the Captains of Industry* (Morgantown, W. Va.: West Virginia University Foundation, 1976).

2. Examples of this perspective may be seen in the following works: Helen Lewis et al., *Colonialism in Modern America: The Appalachian Case* (Boone, N.C.: The Appalachian Consortium Press, 1978); Keith Dix, "Appalachia: Third World Pillage?" in *Appalachia: Social Context Past and Present*, ed. Bruce Ergood and Bruce E. Kuhre (Dubuque, Iowa: Kendall Hunt, 1976), pp. 167-72; and Emil Mazilia, "Economic Imperialism: An Interpretation of Appalachian Underdevelopment," in *Appalachia: Social Context*, pp. 162-67.

3. For a discussion of the transitions taking place during this time period, see Ronald D. Eller, "Industrialization and Social Change in Appalachia, 1880-1930: A Look at the Static Image," in Lewis, *Colonialism in Modern America*, pp. 35-46. For an in-depth case study of one area, see John Gaventa, *Power and Powerlessness in an Appalachian Valley* (Urbana, Ill.: University of Illinois Press, 1980).

4. For the seventy-two counties in the sample, corporate ownership of surface is associated (though not strongly) with the level of known coal reserves, such that the greater the reserves, the greater the corporate ownership (Pearson's R = .368 at the .0001 level of significance) and the greater the corporate ownership of mineral rights (Pearson's R = .369 at the .005 level of significance). Even within the major coal counties (i.e., those with the greatest reserves), corporate ownership increases with the level of coal production (Pearson's R = .437 at the .001 level) for surface ownership, and with the "value added" in mining (Pearson's R = .433 at the .036 level in the case of surface ownership and .468 at the .030 level in the case of mineral rights).

5. For the seventy-two rural counties in the sample, the relationship between the degree of corporate ownership of land and minerals and the percentage of the labor force in mining is significant (Pearson's R = .479 at the .0001 level of significance in the case of surface and .621 at the .0001 level in the case of mineral ownership). This might be expected, because we have already found corporate ownership to be associated with the level of coal reserves. However, even in the case of thirty-seven counties with a high level of reserves, the relationship holds: corporate ownership means a heavy concentration of the labor force in mining (Pearson's R = .580 at the .0002 level in the case of surface ownership and .560 at the .001 level in the mineral ownership).

6. The association between absentee ownership and increased coal production is not a strong one in the case of all absentee (out-of-county and out-of-state) owners. (Pearson's R = .326 at the .052 level). However, it increases in strength when only out-of-state owners are considered (Pearson's R = .450 at the .006 level). This would lend support to the finding that the controllers of the coal production are located in metropolitan centers out of the region.

7. Appalachian Regional Commission, *Capital Resources in the Central Appalachian Region*, Report No. 9 (Washington, D.C.: Checci & Co., 1969).
8. For corporate ownership, the Pearson's R correlation is −.453 at the .005 level. For out-of-state ownership, it is −.486 at the .002 level of significance.
9. The association between out-of-state ownership of land in a county and the number of manufacturing establishments in 1972 is Pearson's R = .357 at the .030 level. In the case of value added in manufacturing, Pearson's R = .441 at the .013 level.
10. For documentation of the effects of boom-town development, see, for instance, Helen Lewis et al., "Coal Productivity and Community: The Impact of the National Energy Plan in the Eastern Coalfields" (paper prepared for the Department of Energy, February 1978).
11. See the report for the U.S. Department of Energy, *Environmentally Based Siting Assessment for Synthetic Fuels Facilities*, January 1980.
12. For thirty-seven coal counties, Pearson's R = .490 at the .002 level of significance.
13. For all absentee ownership, Pearson's R = .405 at the .013 level. For out-of-state ownership, the level rose to .539 at the .001 level.
14. For forty-four counties for which data were available, Pearson's R correlation = .609 at the .0001 level of significance. The level of tourism development was measured as the percentage of service industries in hotels, motels, trailer parks, campgrounds, amusement and recreations, according to the U.S. Bureau of the Census 1972 Census of Services.
15. William Cary, Molly Johnson, Meredith Golden, and Trip Van Noppen, *The Impact of Recreational Development: A Study of Land Ownership, Recreational Development and Local Land Use Planning in the North Carolina Mountain Region* (Durham, N.C.: North Carolina Public Interest Research Group, 1975).
16. See Anita Parlow, "The Land Development Rag," in Lewis, *Colonialism in Modern America*, pp. 177-98, for a discussion of some of these studies.
17. This is the argument of Edgar Bingham, a professor at Emory and Henry College in southwestern Virginia, in his article, "The Impact of Recreational Development in Pioneer Life Styles in Southern Appalachia," in Lewis, *Colonialism in Modern America*, p. 59.
18. As discussed in Parlow, "Land Development Rag," p. 190.
19. For nineteen major tourist counties, the association between corporate and government ownership of land and level of unemployment (1977) is .472 at the .041 level of significance. For concentration of ownership (i.e., large amounts of land controlled by few owners), the strength of this relationship rises even further (Pearson's R = .580 at the .009 level using the Gini concentration coefficient). Given the small number of counties in the sample, both of these relationships are significant.
20. For the nineteen tourist counties, the Pearson's R correlation between degree of corporate ownership of land and per capita income (1974) is −.486 at the .035 level of significance. For percentage of families below the poverty line (1969) the Pearson's correlation is .469 at the .043 level.
21. For instance, for the nineteen tourist counties absentee ownership is negatively related both to total bank deposits (Pearson's R = −.469 at the .003 level and total time deposits (−.468 at the .043 level).

CHAPTER 5

1. Cited in Cynthia Guyer, "Landownership and Inequality in Appalachia," Senior thesis, Oakes College, 1975, p. 77.
2. U.S. Department of Agriculture, Economics, Statistics, and Cooperatives Service, *Status of the Family Farm: Second Annual Report to Congress*, Agricultural Economic Report 434 (Washington, D.C.: U.S. Government Printing Office, 1979), p. 2.

3. The literature analyzing these factors is summarized in Steve Fisher and Mary Harnish, "Losing a Bit of Ourselves: The Decline of the Small Farmer" (paper presented at the 1980 meeting of the Appalachian Studies Conference, East Tennessee State University, Johnson City, Tenn.).

4. Much of this analysis is taken from Dean Pierce, "The Low-Income Farmer: A Reassessment," *Social Work in Appalachia* 3 (1971):7-10. Pierce offers a good, concise summary of the historical development of agriculture in Appalachia. He relies heavily upon the analyses by Harry Caudill, *Night Comes to the Cumberlands* (Boston: Atlantic–Little, Brown, 1962), and Anthony Caruso, *The Appalachian Frontier* (Indianapolis: Bobbs-Merrill, 1959).

5. Caudill, *Night Comes*, pp. 61-65, 71.

6. Ibid., pp. 74-76.

7. Pierce, "Low-Income Farmer," p. 8.

8. See *ALOS*, vol. 2.

9. Malcolm Ross, *Machine Age in the Hills* (New York: Macmillan Co., 1933), p. 84.

10. Roy E. Proctor and T. Kelly White, "Agriculture: A Reassessment," in *The Southern Appalachian Region: A Survey*, ed. Thomas R. Ford (Lexington, Ky.: University of Kentucky Press, 1962), p. 87.

11. Unless otherwise indicated, agricultural data used in these correlations is based upon the 1974 Census of Agriculture. Recognition is given to the possible difficulties of correlating ownership in 1978-79 to these agricultural traits.

12. In the seventy-two rural counties in the sample, the association between percentage of a county corporately-owned and the percentage of land in agriculture is significant (Pearson's $R = -.498$ at the .0001 level). The relationship increases in strength when both corporate and public land are included, rising to Pearson's $R = -.519$ at the .0001 level. Outside of the coalfields, 68 percent of the counties studied have a high degree of agricultural land use, and all of these have a low degree of corporate control. For twenty-two counties outside the coalfields, the Pearson's R correlation between percentage of county held by corporations and government and the degree of agricultural land use is $-.622$ at the .002 level of probability.

13. The correlation (Pearson's R) between the percentage of a county absentee-owned (i.e., by out-of-state and out-of-county owners) and the percent of land used for agriculture is $-.429$ at the .0002 level of probability. For noncoal counties, the negative relationship strengthens to $-.666$ at the .0007 level.

14. The greater the concentration of land the less the percentage of the county used for agriculture. The correlation (Pearson's R) between the Gini coefficient and percentage of county in agriculture is $-.499$ at the .0001 level of probability.

15. For the thirty high agricultural counties for which data was available, the percentage of farmers with other occupations as a principal income source correlates strongly with several landownership patterns, as follows: (a) percentage of county absentee held—Pearson's $R = .380$ at the .038 level of probability; (b) percentage of county in corporate and government ownership—$R = -.451$ at the .012 level of probability; (c) percentage of county in absentee corporate and government ownership—$R = .517$ at the .003 level of probability; and (d) concentration (Gini coefficient)—$R = .723$ at the .0001 level of probability.

16. For the nineteen tourist counties in the sample, the greater the percentage of a county in government ownership, the greater the percentage of farmers with other occupations as principal income source (Pearson's $R = -.597$ at the .009 level of significance). For corporate and government ownership combined, the relationship increases in strength (Pearson's $R = .706$ at the .0007 level).

17. The relationships here are very high, especially for such a small number of counties ($n = 19$). For government ownership, Pearson's $R = .817$ at the .0001 level; for absentee ownership, $R = .734$ at the .0003 level; and for concentration, $R = .603$ at the .006 level, using the Gini coefficient, and .846 at the .0001 level, using the concentration index (see methodology section for description).

18. Coal production is based on 1977 data. For the seventy-two rural counties in the sample, Pearson's R is as follows: the greater the level of coal production, the fewer the farms in a county (−.398 at the .002 level), the less the farm acreage (−.441 at the .0004 level), and the lower the percentage of the county in agricultural use (−.540 at the .0001 level).

19. For thirty-one counties for which data was available, the Pearson's R correlation between the corporate control of mineral rights (expressed as percentage of county surface) and the loss in number of farmers between 1969-1974 is strong (.504 at the .004 level of probability). For corporate control of mineral and surface combined (Index of Resource Control), it is .533 at the .002 level.

20. Pearson's R = −.525 at the .001 level of probability.

21. See *ALOS*, 3:64-99.

22. For thirty agricultural counties, the Pearson's R correlation between the percentage of the county surface owned by corporations and the percentage of the county in agriculture was −.472 at a .008 level of probability. For twenty-two agricultural counties with mineral rights, the correlation between percentage of county underlain by corporately-owned minerals and the percentage of the county in agriculture was −.576 at the .005 level. When the percentage of corporate ownership of surface and mineral acres was combined for twenty-two counties (Index of Resource Control), Pearson's R rose to −.665 at the .0007 level of probability. The correlation between level of absentee ownership and percentage of county in agriculture was similar: −.403 at the .027 level for surface acres; −.527 at the .010 level for mineral acres; and −.656 at the .0007 level for surface and mineral combined.

23. For thirty agricultural counties, the Pearson's R correlation between percentage of the county corporately-owned and the percentage of farmers with other occupations was .450 at the .012 level of probability for surface acres, .524 at the .012 level for mineral acres, and .577 at the .005 level for surface and mineral acres combined.

24. See *ALOS*, 2:71.

25. Walter Goldschmidt, *As You Sow: Three Studies in the Social Consequences of Agribusiness* (Glencoe, Ill.: Free Press, 1947; reprt. Montclair, New Jersey: Allanheld, Osmun & Co., 1978). For a discussion of other studies of these impacts, see particularly the introductory section, "Agriculture and the Social Order," pp. xxiii–liv in the reprint edition.

26. Quoted in Jim Hightower, *Eat Your Heart Out: Food Profiteering in America* (New York: Vintage Books, 1976), p. 158.

27. For further information on these arguments, see Fisher and Harnish, "Losing a Bit of Ourselves," pp. 11-14.

CHAPTER 6

1. West Virginia Governor's Housing Advisory Commission. "Final Report to Governor John D. Rockefeller IV," January 1980, p. 29.

2. President's Commission on Coal, *The American Coal Miner: A Report on Community and Living Conditions in the Coalfields* (Washington, D.C.: U.S. Government Printing Office, 1980), p. 54.

3. Ibid.

4. Unless otherwise indicated, housing data are from the 1970 Census, which was the most recent available at the time of writing.

5. For seventy-two counties, Pearson's R = .490 at the .0001 level of probability.

6. For seventy-two counties, Pearson's R correlation = .419 at the .002 level of probability.

7. For seventy-two counties, Pearson's R correlation = .411 at the .0003 level of probability.

8. For thirty-seven coal counties, Pearson's R correlation = .435 at the .007 level of probability, such that the greater the degree of "unavailable" land (corporate + government + absentee), the greater the degree of overcrowded housing.

9. For twenty-two noncoal counties, Pearson's R = .656 at the .0013 level of probability.

10. For twenty-two noncoal counties, the Pearson's R correlation between percentage of county in corporate ownership and level of overcrowdedness is only .240 at the .283 level of probability. However, for absentee ownership, the correlation is .634 at the .001 level. In these counties, a high degree of government ownership is also associated with overcrowded housing (Pearson's R = .486 at the .030 level of probability).

11. The Pearson's R correlation between level of corporation ownership and overcrowded housing is. .369 at the .025 level; for absentee ownership it is .511 at the .001 level of probability.

12. For thirty-seven coal counties, Pearson's R = .331 at the .045 level of probability.

13. For thirty-five coal counties for which data on mineral rights were available, Pearson's R = .457 at the .006 level of probability.

14. Pearson's R = .425 at the .01 level of probability.

15. This case, including quotations, is taken from John Gaventa, *Power and Powerlessness in an Appalachian Valley* (Urbana, Ill.: University of Illinois Press, 1980), pp. 125-35.

16. Further documentation of the capital outflow patterns may be found in Appalachian Regional Commission, *Capital Resources in the Central Appalachian Region,* Report No. 9 (Washington, D.C.: Checci & Co., 1969).

17. For thirty-seven coal counties, Pearson's R correlation = -.318 at the .055 level of probability.

18. Data developed by Errol Hess, Rural Area Development Association, Scott County, Virginia.

19. For nineteen tourist/recreation counties, Pearson's R correlation = .557 at the .013 level of probability.

20. For thirty-seven coal counties, the Pearson's R correlation between percent of corporate-controlled land in a county and percent of owner-occupied dwellings is -.488 at the .002 level of probability.

21. Quote in *Appalachia* 12 (Fall 1978): 12.

22. U.S., Department of Energy, "Synthetic Fuels and the Environment: An Environmental and Regulatory Analysis," DOE 9V-0087 (June 1980), p. 147.

CHAPTER 7

1. *ALOS,* 7:162.
2. Wheeling, West Virginia, *News Register,* 3 December 1980.
3. *ALOS,* 7:52.
4. Ibid., p. 54.
5. *ALOS,* 5:4.
6. *ALOS,* 6:179.
7. Louisville, Kentucky, *Courier Journal,* 1 January 1980.
8. Knoxville, Tennessee, *News Sentinel,* 26 November 1979.
9. *ALOS,* 3:102.
10. *ALOS,* 7:52.
11. *ALOS,* 6:147.
12. Kingsport, Tennessee, *Times,* 3 April 1980.
13. *ALOS,* 5:26.

14. Ibid., p. 83.
15. *ALOS,* 2:42.
16. John Gaventa, "Review of Leasing Activity in Twelve Counties Along the Tennessee-Tombigbee Waterway, unpublished report, 1978.
17. *Wall Street Journal,* 12 December 1980.
18. Ibid.
19. *Dun's Review,* May 1980; and *Washington Post,* 8 October 1980.
20. John M. Dennison and Walter W. Wheeler, "Stratigraphy of Precambrian through Cretaceous Strata in Southeastern United States and Their Potential as Uranium Host Rocks," *Southeastern Geology,* Special Publication No. 5 (July 1975).
21. See, for example, the account of uranium mining and milling in chapter 2 of *Environmental Ethics* by Science Action Coalition with Albert J. Fritsch (New York: Andros Books, 1980).
22. Reported in *Roanoke Times and World News,* 30 November 1980.
23. Breathitt, Johnson, Knox, Laurel, Martin, Perry, and Bell counties.
24. Charles R. Collier et al., "Influences of Strip Mining in the Hydrologic Environment of Parts of Beaver Creek Basin, Kentucky, 1955-59," U.S. Geological Survey, Survey Professional Paper 627 (A-C), 1964.
25. Reported in Hong-Shong Tung, "Impacts of Contour Coal Mining on Stream Flow: A Case Study of the New River Watershed, Tennessee" (Ph.D. dissertation, University of Tennessee, Knoxville, Tenn., 1975).
26. See, for example, Kenneth L. Dyer and Willie R. Curtis, "Effects of Strip Mining on Water Quality in Small Streams in Eastern Kentucky, 1967-1975," U.S. Department of Agriculture, Forest Service Research Paper NE-372, 1977.
27. Willie R. Curtis, "The Curtis Report," *Green Lands,* West Virginia Surface Mining and Reclamation Association, Summer 1977, p. 21.
28. Tug Valley Recovery Center, *A Clear and Imminent Danger: The Case for Designating the Tug Fork Watershed Unsuitable for Strip Mining* (October 1980).
29. U.S., Environmental Protection Agency, *Erosion and Sediment Control: Surface Mining in the Eastern United States,* Planning Volume, p. 5.
30. Stanford Research Institute, *A Study of Surface Coal Mining in West Virginia* (1972), p. 56.
31. See above notes 8 and 10.
32. Willie R. Curtis et al., "Fluvial Sediment Study of Fishtrap and Dewey Lakes Drainage Basins, Kentucky-Virginia," U.S. Geological Survey Water Resources Investigations 77-123 (March 1978), prepared in conjunction with U.S. Army Corps of Engineers.
33. Kentucky Department of Natural Resources, *The Floods of April 1977* (n.d.), p. 157.
34. U.S. Army Corps of Engineers, Huntington District, *The 1977 Tug Fork Valley Flood* (n.d.), pp. 6-11.
35. Tug Valley Recovery Center (see above note 28).
36. Extensively documented in water quality studies by Dyer and Curtis (See above note 26).
37. Commonwealth of Kentucky, Department for Natural Resources and Environmental Protection, Division of Water Quality, *1978 Kentucky Water Quality 305(b), Report to Congress.*
38. Tennessee Valley Authority, *Is the Water Getting Cleaner?* (November 1980).
39. Southwest Virginia 208 Planning Agency, *Southwest Virginia 208 Plan* (Duffield, Va., June 1978), p. 206
40. Tug Valley Recovery Center, *Clear and Imminent Danger,* p. 43.
41. Quoted in Kingsport, Tennessee, *Times News,* 3 April 1980.

42. U.S. Department of Energy, Assistant Secretary for the Environment, Office of Technology Impacts, *Environmentally Based Siting Assessment for Synthetic Liquid Fuels Facilities,* Draft Final Report DOE/EV/10287 (December 1979), p. IX-1.

43. Highlander Research and Education Center, *We're Tired of Being Guinea Pigs!* (New Market, Tenn., Summer 1980).

44. Reported in *Coalfield Progress,* 13 March 1980.

45. Surveyed in H. M. Braunstein, E. D. Copenhaver, and H. A. Pfuderer, eds., *Environmental, Health, and Control Aspects of Coal Conversion* (Oak Ridge, Tennessee: Oak Ridge National Laboratory, prepared for Energy Research and Development Administration, April 1977), vols. 1 and 2.

46. *Eastern Oil Shale: A New Resource for Clean Fuels* (Chicago: Institute for Gas Technology, 1980).

47. Reported in *Synfuels,* 16 May 1980.

48. Louisville *Courier Journal,* 16 July and 5 August 1980.

49. Conversation with Tim Murphy, director, Kentucky Rivers Coalition, 13 July 1982.

50. Reported in Louisville *Courier Journal,* 5 May 1980.

51. See special energy section in *Fortune,* 24 September 1979.

52. *Wall Street Journal,* 12 December 1980, p. 56.

53. Virginia Department of Labor and Industry, quoted in *Kingsport* (Tenn.) *Times,* 23 September 1979.

54. U.S. Forest Service, "Environmental Assessment for Issuance of Oil and Gas Leases," 13 August 1980.

55. Coalition of Appalachian Energy Consumers, *Communities Battle Pumped Storage,* pamphlet (Dungannon, Va.: n.d.).

56. Reported in, for example: Robert O. Becker, "Brain Pollution," *Psychology Today,* February 1979, p. 124; and Sierra Club Legal Defense Fund, "Exhibit Summary of Reported Data Relating to Effects of 765 KU Transmission Lines," submitted to Federal Energy Regulatory Commission, (FERC), n.d.

57. *Mountain Life and Work,* January 1981, pp. 11-14.

CHAPTER 8

1. Bernard J. Nieman, ed., *Land Record Systems Can And Should Be Modernized: Selected Papers from American Institute of Planners 60th Anniversary Conference* (Madison, Wisc.: Institute for Environmental Studies, 1980).

2. David Liden, "Taxing Companies Properly Could Net State $150 Million," *Charleston* (W.Va.) *Gazette,* Saturday, July 10, 1982, p. 5A.

3. Ibid.

4. *Struggling For Tax Justice in the Mountains* (Lovely, Ky.: Kentucky Fair Tax Coalition June, 1982, p. 10.

5. David Liden, "Updating Resource Taxes To Aid Schools," *Charleston* (W.Va.) *Gazette,* July 9, 1982, p. 7A.

6. Charles C. Geisler, "Toward Reform in Land Reform: Coupling Local Control and Social Control" (paper presented at the annual meeting of the American Sociological Association, Boston, 1979).

7. "Opinion and Findings of Fact," May 11, 1982, issued by Arthur M. Recht, Special Judge, Circuit Court, Kanawha County, West Virginia.

APPENDIX 2

1. Janet Smith, David Ostendorf, and Mike Schechtman, *Who's Mining The Farm?* (Herrin, Ill.: Illinois South Project, 1978).

2. Nancy R. Bain, "The Impact of Absentee Land Ownership in a Rural Area: A Case Study in Southeast Ohio" (unpublished paper, Department of Geography, Ohio University, Athens, Ohio, 1977); also Ralph W. Kline, *The Extent and Characteristics of Absentee Land Ownership Within A Five County Region of Southeastern Ohio* (M.A. Thesis, Ohio University, Athens, Ohio, 1977).

3. Richard M. Kirby, "Kentucky Coal: Owners, Taxes, Profits" (paper prepared for the Appalachian Volunteers, Prestonsburg, Ky., 1969). See also excerpt in *Appalachian Lookout* (Oct. 1969), pp. 19-27.

4. Joey Childers, "Harlan County Land Ownership" (unpublished paper, University of Kentucky Appalachian Center, Lexington, Ky., 1978).

5. William Carey, Molly Johnson, Meredith Golden, and Trip Van Noppen, *The Impact of Recreational Development: A Study of Land Ownership, Recreational Development and Local Land Use Planning in the North Carolina Mountain Region* (Durham, N.C.: North Carolina Public Interest Research Group, 1975).

6. John Gaventa, Ellen Ormond, and Bob Thompson, "Coal, Taxation, and Tennessee Royalists" (Nashville, Tenn.: Vanderbilt Student Health Coalition, 1971), photocopy.

7. J. Davitt McAteer, *Coal Mine Health and Safety: The Case of West Virginia* (New York: Praeger Publishers, 1973), pp. 140-80.

8. Tom D. Miller, *Who Owns West Virginia?* (Huntington, West Va.: The Huntington Publishing Company, 1975).

9. John Gaventa and Billy D. Horton, "A Citizens' Research Project in Appalachia, U.S.A.," *Convergence XIV*, 3 (1981): 30-42; also, John Gaventa and Bill Horton, "Digging the Facts," *Southern Exposure* X, 1 (January-February, 1982): 34-38. "For further information on this method, see: International Council of Adult Education, *Convergence* XIV 3(1981) and Thord Erasmie and Jan deVries, eds., *Research for the People—Research by the People: Selected Papers from the International Forum on Participatory Research in Ljubljana, Yugoslavia, 1980* (Sweden: Linkoping University Department of Education, 1981).

10. Bill Horton, "How To Find The Facts," *Southern Exposure* X, 1 (January-February, 1982): 42-43.

11. Bain, "Impact of Absentee Land Ownership"; Kline, "Extent and Characteristics of Absentee Land Ownership."

12. Walter Goldschmidt, *As You Sow: Three Studies in the Social Consequences of Agribusiness* (Glencoe, Ill.: Free Press, 1947: repr, Montclair, N.J.: Allanheld, Osmun, and Company, 1978); Kirby "Kentucky Coal"; Cary, et. al., *The Impact of Recreational Development*; and Miller, *Who Owns West Virginia?*

13. Gene Wunderlich, "Concentration of Land Ownership," *Journal of Farm Economics,* SL(5) (December, 1958): pp. 1887-93).

INDEX

absentee ownership, 8, 12, 15, 20, 88, 95; deters land improvement, 8; hinders economic diversification, 68; in tourism counties, 76, 79; makes land unavailable, 100; direct impacts of, 141
acid mine drainage, 123
active leases, 165
Addington Oil Company, 128, 129
ad valorem property tax, 52
agribusiness, 3, 80
agricultural land assessment, 145
agriculture, decline of, 3, 4, 80. *See also* Grayson County, Virginia
Alabama, 47-49, 146, 167
Alabamians for Tax Relief Committee, 47, 48
Alaska, government ownership of, 2
Alleghany Power System, plans for pumped storage, 134
American Association Ltd, 30, 96, 103, 104
American Syn-Crude, 129
Amherst Coal Company, 120, 121
AMOCO, 38
Appalachian Alliance: formation of, vi; role in landsurvey, 159
Appalachian Landownership Study, vi, vii; citizens input in, viii; as basis for community action, ix; funding of, x; need for, 13; practical application of, 146; publications concerning, 162; states included in, 163
Appalachian Land Ownership Task Force, vi, xi, 14, 15; composition of, 159
Appalachian Power Company (APCO), proposed pump storage, 133, 134
Appalachian Regional Commission (ARC), vi, vii, x, 67, 77, 170
Appalachian State University, 76
Appalachian Studies Conference, vi
Army Corps of Engineers. *See* United States Army Corps of Engineers

Big Sandy Development District, housing plans for, 96
black landowners in the south, 4
boom and bust cycles, 70, 71
Braden's Flats, Tennessee, 110
Brimstone Company, 29
Brumley Gap, Virginia, pumped storage protest in, 133
Buchanan County, Virginia, 68
Buncombe County, North Carolina, 36
Bureau of Public Works in West Virginia, 164

California, landownership in, 17
Cambell County, Tennessee, 96, 108, 112, 125
case study counties: selection of, 169; in state reports, 175
cash crops, 93
Cass Scenic Railroad, 75
Catlettsburg, Kentucky, 125
Census of Governments (1977), 41, 108, 170, 173, 174
Census of Selected Service Industries, 174
Center for Appalachian Studies, ix
chromium, 118
citizen action, recommendations for, 138
citizens protest: strip mining, 116; oil shale development, 129; pumped storage facilities, 133, 134
coal, as source of synthetic fuel, 119
coal, future of, 70
coal camps dependent on cash economy, 82
coal companies: property tax reduction for, 47; national ownership of, 114, 115
Coal Creek Mining and Manufacturing, 110
coal development: effects on small farms, 84; in agricultural counties, 92; effects on housing market, 96; in fringe areas, 116. *See also* energy crisis
coal industry, economic effects of, 64. *See*

also Cumberland Gap, Kentucky; Harlan County, Kentucky
Coalition of Appalachian Electric Consumers (CAEC), 134
coal leasing, increase of, 116
coal mining, in national forest, 37, 38
coal production: expected increases in, 53; from deep mine to strip, 120
coal reserves, present value of, 53
colonialism, 64
commercial property, taxation of, 47, 161
community development, 9
company housing, 103, 104, 109. *See also* Dingess Rum Coal Company
company towns, 32, 103, 104
Concerned Citizens of Martin County (Kentucky), 124
Consolidation Coal Company, 72
Continental Oil Company, 114
copper, 118
corporate headquarters located outside region, 28
corporate land use, 115
corporate ownership, 69, 161, 162; in coalfields, 23; effects of, on community development, 27; effects of, on economic development, 65; of farm lands, 80
corporate planning, lack of public knowledge of, 72
corporate reports, where available, 162
corporate tax rates, 43
Cotiga Development Company, 29
Council of State Governments, report on property taxation, 56
county services, lack of revenue for, 59
county tax rolls, as source of data, 158
Cumberland County, Tennessee, 77
Cumberland Gap, Kentucky, 30
Cumberland Plateau, coal speculation in, 116

dams, hydroelectric, 8. *See also* United States Army Corps of Engineers; Tennessee Valley Authority
Daniel Boone National Forest, strip mining in, 38
data, analysis of, 170-72
deed/lease books, as source of corporate ownership data, 161, 162
Dekalb County, Alabama, 34, 91, 93
Department of Energy, 31, 126-28
Dingess Rum Coal Company, 109, 110
diversified industry, lack of, in coal economy, 68
Dungannon, Virginia, proposed synfuels site, 126
Dynalectron, Inc., 125

Eastern Gas Shale Project, 31, 128
Eastern Overthrust Belt, oil and gas leasing in, 31, 117, 132, 143
economic development in tourism counties, 76
economic diversification, 67-69, 79. *See also* landownership patterns
economic instability in boom and bust coal, 64
economic underdevelopment, characteristic of region, 64
economy: based on tourism, 36; farm, 89; from agriculture to coal, 92
Elkhorn City Land Company, 97
eminent domain, 6, 146
energy conglomerates, 143
energy corporations, list of, 116, 117
energy crisis, 4, 118
energy development: impacts of, on housing, 100, 101; environmental costs of, 119; future effects of, 135
energy expansion, in periphery areas, 34
energy independence, push for, 126
energy market, 72
Environmental Protection Agency, 122

fair market value, 45, 52, 139
Farmers' Home Administration (FHA), 97, 106
farm income, 86, 92
farmland: loss of, 8; sold for back taxes, 83; used in recreational development, 88
farmland exemptions, 46, 47
federal and state supplements to tax-poor counties, 59
federal government, largest land holder, 13
federal holdings tax exempt, 13
federal ownership of recreational areas, 75. *See also* Swain County, North Carolina
Federation of Appalachian Housing Enterprises, 142
field researchers, training of, 160
flood (1977), 108, 122
flooding, effects on farmland, 82
formaldehyde gas, in mobile homes, 111
forest resources, exhaustion of, 74-76
Forest Service. *See* United States Forest Service
Friends of the Little Kanawa in West Virginia, 143

Gary, West Virginia, 32
gas and oil wells, now producing, 130, 131
gasification plant, 125
government housing programs, 107
government ownership: of western coal, 5; in tourism counties, 75, 88; of farmland,

Index

84; sources of data, 164. *See also* federal ownership
government policies on agriculture, 81
Governor's Housing Advisory Commission, of West Virginia, 95
Grandfather Mountain, North Carolina, uranium exploration on, 33, 118
Grayson County, Virginia, 93
Grayson Highlands State Park, 84; effect on local economy, 76
Great Smoky Mountains National Park, creation of, 74
Greenwood Mining Company, 38
Gulf Oil Company, 113

Harlan County, Kentucky: ownership of, 29; poverty levels in, 67; housing crisis in, 96, 104-106, flooding in, 123
Harvard University owns oil and gas rights, 38
Haskold formula of taxing coal in place, 50
housing: shortage of, 69, unavailability of land for, 78; age of, 101. *See also* energy development; mobile homes
housing and urban development, 97, 106, 107, 108
Housing Census, 170
housing crisis, contributing factors to, 95
housing development, financial barriers to, 105, 106
housing units, funded by government, 111, 112
Huber Corporation (J.M.), 22
Huntington Herald, 159

Illinois, land ownership study in, 6
impact mitigation, 138
income in nonagricultural counties, 92
individual ownership, primarily absentee, 38
industry, early development of, 65
Institute of Gas Technology, report on oil shale reserves, 128
Island Creek Coal Company, 120

Jefferson National Forest: initial purchase of, 82; uranium exploration in, 118
Jenkins, Kentucky, 32

Kentenia Corporation, 29
Kentucky, mineral taxation in, 48, 49
Kentucky coalfields, ownership of, 10
Kentucky Consumer Protection Division, 32
Kentucky Department of Revenue, 137, 167

Kentucky Fair Tax Coalition, for unmined minerals tax, 140, 147
Kentucky River coal, 25
Kentucky River Coal and Coke, 29
Kentucky Water Quality Report to Congress, 123
Knobs area of central Kentucky, oil shale leasing in, 34, 128
Koppers Company, 31, 112

labor force: involved in mining, 66; involved in manufacturing, 68
land: removed from market, 69; unavailability of, 69, 96, 142; competition for, 85; cultural value, 94; shortage of, in housing industry, 97, 98
landowners, types of, 173
landownership: concentration of, 3, 17, 120; of energy lands, 6; corporate domination of, 14; effects of, on housing, 105; categories of, 164; variations of, 176. *See also* Appalachian Landownership Study
landownership census system, need for, 137
landownership data, deficiency of, 2
landownership information, control of, 136
landownership patterns: effect on economic diversification, 68, 69; impacts of, on quality of life, 170
land reform, 2, 138, 141, 146
land use, categories of, 166
land values, in recreation and tourism counties, 84
large tracts, underassessment of, 46
LENO-WISCO, planning board of Virginia, 70
lessor/lessee index, 162
lignite resources, use of, 117
Lincoln County, West Virginia, 25, 62, 132
local capital, unavailable in coal regions, 70
local governments, fiscal crisis in, 41
local ownership in agricultural counties, 39, 40
Logan County, West Virginia, 68-70, 109

Martiki Coal Company, 43
Martin County, Kentucky, 43, 59, 60, 69, 96, 97, 115
migration, 65, 72-74
Mineral County, West Virginia, 19, 117
mineral ownership: separate from surface, 5; lack of documentation of, 18, 167; corporate control of, 25; by absentee holders, 30
mineral rights: corporate ownership of, 16; taxation of, 25, 41, 139; ownership of, 37;

with land titles, 97; degrees of ownership of, 143
mineral taxation, 48-50
Mingo County, West Virginia, vi, 89, 120, 122, 124
Mitchell County, North Carolina, mining activities in, 33
mobile homes, 109, 111. *See also* formaldehyde gas
Morgan County, Tennessee, 57
mountain agriculture, 81
Mountain Parkway, impact on local communities, 37
mountaintop removal, 34, 35
Mount Rogers Recreation Area: tourism in, 37; effects of, on local economy, 77; purchase of, 84

National Coal Lessers' Association, 27
National Energy Act, call for coal study, 6
national forest lands: in southern mountains, 13; involvement in tourism, 75; oil and gas exploration in, 132; taxation of, 140, 141
Nature Reserves Commission, 123
New River studies by University of Tennessee, 121
North Carolina, 12, 36, 49, 75
North Carolina Public Interest Research Group (NC PIRG), 12, 75, 158

Oak Ridge, Tennessee, nuclear processing in, 35
Occidental Petroleum, 34, 113
Office of Surface Mining (OSM), 125
oil and gas companies, 120, 121; list of, 117; speculation in coal, 30, 114; syn-fuel production, 31
oil and gas leasing: in southwest Virginia 38; in western North Carolina, 38; increase of, 117, 118; in national forest, 132. *See also* Eastern Overthrust Belt
oil drilling, environmental hazards of, 132
oil shale, 31, 129, 130
oil strikes, 117
orphan lands, 124
overcrowding in coal counties, 99

paper and pulp industry, 3
parent companies, identification of, 165
Payments in Lieu of Taxes Act for federal lands, 56
Pike County, Kentucky, 67, 70; lack of alternative industry in, 69; overcrowding in, 111
Pittston Coal, 11
Pocahontas Kentucky Corporation, 115

population growth in coal fields, 71
poverty, 10; in coalfield counties, 67, 92; in recreation areas, 79
property assessment, 52, 138, 172
property tax, 41, 59, 85
Presidents Coal Commission, 5
private land, purchase by government, 6
public education, 60-63
public lands, 6; recreational development in, 13; private mining of, 38
public services, 141
pumped storage facilities: resistance to, 133-35; environmental impacts of, 134. *See also* Appalachian Power Company; Brumley Gap, Virginia

railroads: ownership of coal reserves by, 33; role of, in industrialization, 65, 74
Randolph County, West Virginia, 75, 132
recreational development, side effects of, 12, 75
recreation and tourism: demands for land, 6; impact on local culture, 89. *See also* Swain County, North Carolina
residency: impact on land use, 165; identification of, 166
residential development of farm lands, 84
resource independence, 118
revenue, loss of, 140
river systems, 127

Save Our Cumberland Mountains (SOCM), 91, 120, 146
school systems in financial crisis, 60-63
Scott County, Virginia, 72, 116
second home development, 97-99; environmental impacts of, 7; near public lands, 12; pressure of, on farm land, 87
Securities and Exchange Commission (SEC), 162
services infrastructure, lack of, 68
severance tax generates revenue, 52
sewage services to rural residents, 107, 108
Shelby County, Alabama, 93
single industry economy, 72
small farm: concentration of, in Appalachia, 8; economic stability of, 9, 93; decline of, 83
social justice through land reform, 147
Soil Conservation Service, 115
Southern Forest Institute, 32
Standard Oil Company, 30, 114
statistics, coding of, 163
strip mining: impact on flood plain, vi, 109, 121,123; silting of farm land from, 91; increase of, 120; environmental effects of, 121

Index

substandard housing, 95
surface and mineral acreage, combined ownership of, 176, 177
surface lands valued low, 42
surface owners, rights over mineral owner, 91, 144
survey counties, selection of, 162, 163
Swain County, North Carolina: absentee ownership of, 19, 22, 36, 87; revenue crisis in, 57; public education in, 61; tourism in, 74; unemployment rates in, 77
synthetic-fuel development, 125-27

taxation: of unmined minerals, 5, 16, 41, 43, 49, 50, 52-55, 139, 140; based on fair cash value, 15; of rural lands, 41, 43; poor property assessment in, 42; of surface acreage, 42, 43; of absentee lands, 43-45; according to land use, 46; relief package for, 47; agricultural exemption in, 47, 145; severance, 48; and property valuation assessors, 55; in coal counties, 55, 69; of state and federal lands, 59, 140; lack of standardization in, 139. *See also* Kentucky Fair Tax Coalition; revenue crisis
tax billing, 161, 165, 173
tax rolls, used in documenting ownership, vii, 160, 168
Tennessee, 11, 35, 50, 57, 116, 117, 131, 144, 158
Tennessee Valley Authority (TVA): floods farm land, 8; ownership of public lands, 12; acreage owned by, 35; seeks waterways for energy development, 36; and syn-fuel development, 125, 126; founding of, 133
Third World Countries, resource extraction in, 118
timber industry, 4, 32; early history of, 74, 75
toxic metals in oil shale mining, 130
toxic minerals produced by strip mining, 123

Tucker County, West Virginia, 134
Tug Valley Recovery Center, 122-24
Tweetsie Railroad, 75

U.S. Steel Corporation, 10
United States Army Corp of Engineers, 12, 13, 36, 116
United States Department of Energy, 35
United States Department of Interior, 35
United States Forest Service, 35-37, 121
United States Geological Survey, 121, 122
underemployment in tourist industry, 78
unemployment rates in tourism-based counties, 77
University of Kentucky studies coal counties, 158
utilities, taxation of, 164
uranium, 118

Virginia, 50, 57, 71-73, 81, 110, 111, 120, 124, 131, 147

Wall Street Journal owns oil rights, 117
Walker County, Alabama, 61, 91, 92, 97, 124
War on Poverty, 10, 59
Watauga County, North Carolina, 74, 76-78; increase in land values in, 88; second home development in, 98
water supplies: use in energy development, 119; in mined areas, 124
western North Carolina, population increase in, 76. *See also* Swain County, North Carolina
West Virginia, 10-11, 25, 66, 74, 104, 112, 158-59; develops unmined coal tax, 51-53, 139-40, 165
Wise County, Virginia, 71-73, 81, 110-11, 120
Wythe County, Virginia, 57

zoning boards, development of, 145, 146
zoning for land use, 144

Library of Congress Cataloging in Publication Data
Main entry under title:
Who owns Appalachia?
   "From the survey of the Appalachian Land Ownership Task Force."
   Bibliography: p.
   Includes index.
   1. Land use, Rural—Appalachian Mountains Region. 2. Land tenure—Appalachian Mountains Region. 3. Mineral industries—Taxation—Appalachian Mountains Region. 4. Real property tax—Appalachian Mountains Region. 5. Local finance—Appalachian Mountains Region. 6. Appalachian Mountains Region—Economic conditions. 7. Appalachian Mountains Region—Social conditions.
I. Appalachian Land Ownership Task Force.
HD210.A66W48   1983   333.3'0974   82-40173
ISBN 978-0-8131-5096-3

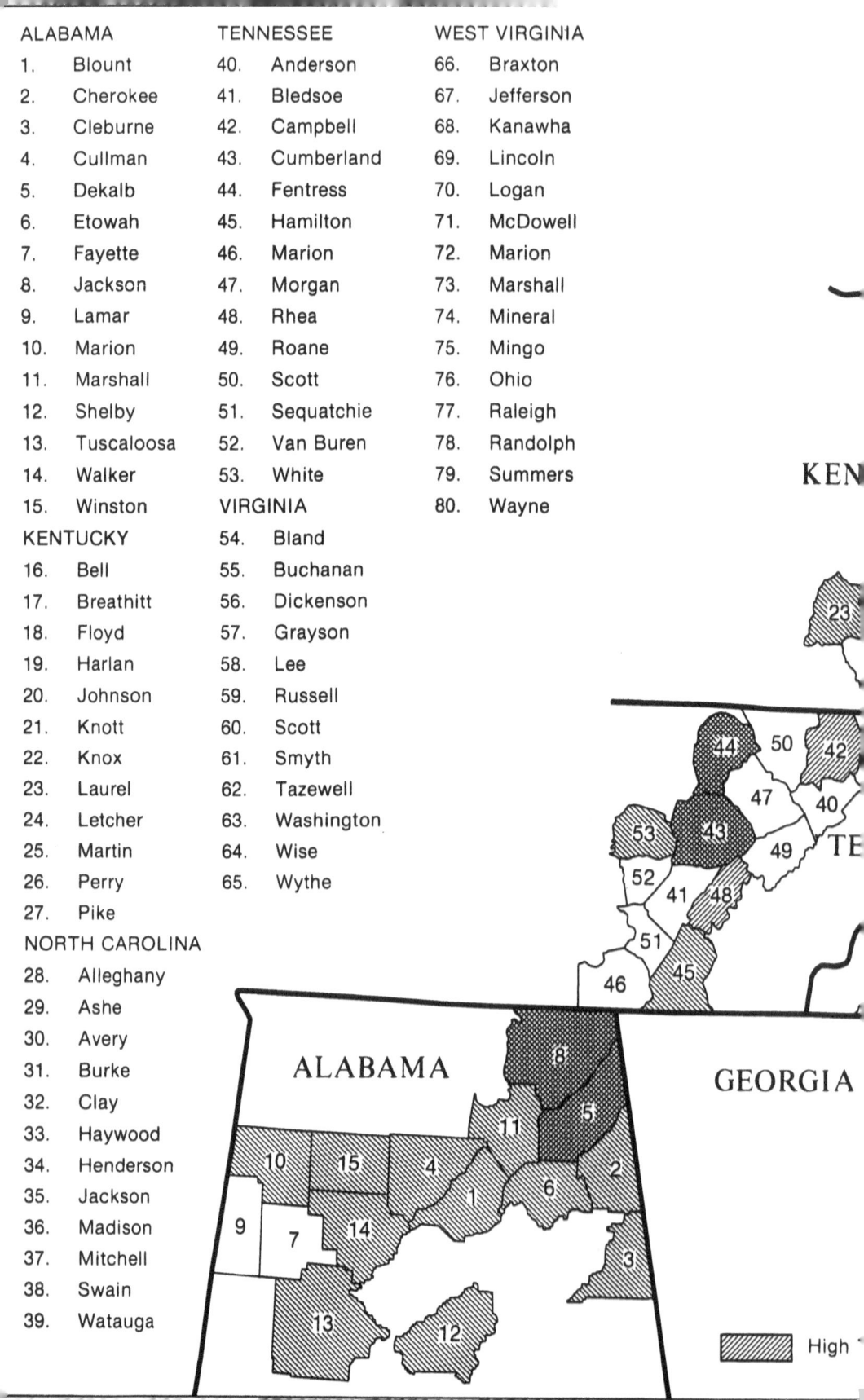

ALABAMA		TENNESSEE		WEST VIRGINIA	
1.	Blount	40.	Anderson	66.	Braxton
2.	Cherokee	41.	Bledsoe	67.	Jefferson
3.	Cleburne	42.	Campbell	68.	Kanawha
4.	Cullman	43.	Cumberland	69.	Lincoln
5.	Dekalb	44.	Fentress	70.	Logan
6.	Etowah	45.	Hamilton	71.	McDowell
7.	Fayette	46.	Marion	72.	Marion
8.	Jackson	47.	Morgan	73.	Marshall
9.	Lamar	48.	Rhea	74.	Mineral
10.	Marion	49.	Roane	75.	Mingo
11.	Marshall	50.	Scott	76.	Ohio
12.	Shelby	51.	Sequatchie	77.	Raleigh
13.	Tuscaloosa	52.	Van Buren	78.	Randolph
14.	Walker	53.	White	79.	Summers
15.	Winston	**VIRGINIA**		80.	Wayne
KENTUCKY		54.	Bland		
16.	Bell	55.	Buchanan		
17.	Breathitt	56.	Dickenson		
18.	Floyd	57.	Grayson		
19.	Harlan	58.	Lee		
20.	Johnson	59.	Russell		
21.	Knott	60.	Scott		
22.	Knox	61.	Smyth		
23.	Laurel	62.	Tazewell		
24.	Letcher	63.	Washington		
25.	Martin	64.	Wise		
26.	Perry	65.	Wythe		
27.	Pike				
NORTH CAROLINA					
28.	Alleghany				
29.	Ashe				
30.	Avery				
31.	Burke				
32.	Clay				
33.	Haywood				
34.	Henderson				
35.	Jackson				
36.	Madison				
37.	Mitchell				
38.	Swain				
39.	Watauga				

www.ingramcontent.com/pod-product-compliance
Lightning Source LLC
Chambersburg PA
CBHW022054160426
43198CB00008B/225